The Female Body in Control

HOW THE CONTROL MECHANISMS IN A WOMAN'S PHYSIOLOGY MAKE HER SPECIAL

Mary Jean Wallace Paxton

Illustrations by Ruth Anne Kocour

A SPECTRUM BOOK

PRENTICE-HALL, INC., *Englewood Cliffs, New Jersey 07632*

Library of Congress Cataloging in Publication Data

PAXTON, MARY JEAN WALLACE.
The female body in control.

(A Spectrum Book)
Includes bibliographies and index.
1. Women—Health and hygiene. 2. Women—Physiology.
3. Biological control systems. I. Title.
RG121.P35 613'04244 81-8682
AACR2

ISBN 0-13-314104-7

ISBN 0-13-314096-2 {PBK.}

A SPECTRUM BOOK. Printed in the United States of America.

10 9 8 7 6 5 4 3 2 1

Editorial/production supervision and interior design by Carol Smith
Cover design by Jeannette Jacobs
Cover illustration by Mona Mark
Manufacturing buyer: Cathie Lenard

PRENTICE-HALL INTERNATIONAL, INC., *London*
PRENTICE-HALL OF AUSTRALIA PTY. LIMITED, *Sydney*
PRENTICE-HALL OF CANADA, LTD., *Toronto*
PRENTICE-HALL OF INDIA PRIVATE LIMITED, *New Delhi*
PRENTICE-HALL OF JAPAN, INC., *Tokyo*
PRENTICE-HALL OF SOUTHEAST ASIA PTE. LTD., *Singapore*
WHITEHALL BOOKS LIMITED, *Wellington, New Zealand*

This book is dedicated to the memory of
Ruth Johnson Wallace, Robert Gerard Haagens,
and John Joseph Cavanaugh, C.S.C.

Mary Jean Wallace Paxton, Ph.D., has taught general biology, anatomy, physiology, and endocrinology at Jacksonville State University, Jacksonville, Alabama; St. Mary's College, Notre Dame; and Rhode Island College, Providence. She has published numerous professional papers in the field of reproductive physiology, and is currently working in research and development for a commercial laboratory in California.

Contents

Preface

We do not lack information about the structure and functions of our bodies. The popular media supplement scientific breakthroughs with more and more facts about the way our bodies operate in health and disease.

What we may lack is an organized frame of reference upon which to build a personal understanding of what happens to us during *adolescence*, while we are *taking the Pill* to avoid pregnancy, during *pregnancy*, or when we enter the *menopause*. This frame of reference should include a knowledge of the factors that control our development (*genes and chromosomes*) and of those systems (nervous and endocrine) that normally regulate our basic body functions.

Our development as genetic females from *puberty* through the *menopause* is regulated by these control systems. *Birth control*, *VD control*, *weight control*, and the *control of alcohol consumption* may be necessary at various times in our lives. We need to know how these factors affect our bodies, how they may interfere with its *control mechanisms*.

Cancer is a disease in which cells grow out of control. It is a disease which may claim the lives of one out of every five American women. We can rarely control the incidence of cancer in our lives. However, we can deal with cancer more effectively if we know what it is and if we use the weapons we have in our hands to oppose it.

ACKNOWLEDGMENTS In organizing information from many sources into a framework for building an understanding of women's physiology, the following people have provided valuable assistance: Dorothy Klingele (Chapter 2); Charles Hoff (Chapter 3); Kenyon S. Tweedell (Chapter 8); and Rosemary Mainland, Sumner Thompson, and Kwei-Hay Wong (Chapter 9). Marian T. Kley contributed her extensive experience in lecturing to medical students on alcoholism and her work as an alcoholism counsellor for the state of California to write Chapter 10. In addition to writing Chapter 11, Carol Ann Dyer provided invaluable editorial assistance and advice on the first nine chapters.

I also acknowledge with gratitude the assistance of Deborah Beaudoin, Rachel Jones, and Barbara Richey, who helped in gathering data; Cindy McCahill, who typed the manuscript; and Jean Carlson, Lois Patterson, and David G. Paxton, who furnished support, moral and otherwise

The Female Body in Control

1 Control Systems

Fear of the unknown has a paralyzing effect. It inhibits our positive efforts to form new relationships, to attempt new approaches to old problems, or to venture into previously unexplored areas. When we are afraid it is harder to grow as persons or to function as free human beings.

For some of us, the functions of our own bodies are feared because they are unknown, at least to us. We are vaguely aware that scientists "out there someplace" have made and are making impressive advances in their knowledge of the mechanisms that control body functions. From time to time the media bring us the latest breakthroughs of modern science: heart transplants, test-tube babies, sperm banks, the wonders of the brain. We are, of course, impressed, but yet, many doubts and questions remain. We may still be fearful of changes in our own bodies, changes that occur on different time scales: the monthly changes of the ovulatory (menstrual) cycle, the continuing changes of a nine-month pregnancy, the developmental changes that characterize puberty and the menopause.

We have, for the most part, survived these changes, even though we didn't always understand what was happening to us. Most of us have struggled through the biological transition from childhood to adulthood known as puberty. Many of us have become or will become mothers, have had or will have some form of gynecological cancer, have had or will have one of the venereal diseases. We shall all, if we live long enough, go through the menopause and grow old. If we take estrogens to make the aging process less debilitating, we wonder what risks to our bodies may be encountered during this form of therapy. We may have problems controlling our intake of food or alcohol, and thus in maintaining the physical fitness that would enable us to be more creative and productive—more *free* to know and to do whatever is paramount in our individual lives, and to take charge of important facets of our own health and well-being.

We can begin by understanding the systems that control the development and functioning of our bodies. It is the aim of this initial chapter to introduce general concepts that can be used as a basis for understanding how *female* bodies operate under various conditions. These general concepts will be specified in subsequent chapters; for example, the section of Chapter 4 on sexual response will describe orgasm (climax) in men and women as a spinal reflex with inputs from higher centers in the brain. This present chapter will explain what a spinal reflex is in terms of nervous control systems. The ovulatory cycle will be explained in Chapter 4 in terms of the hormones that regulate it; this chapter will define general characteristics of hormonal (endocrine) control systems.

Both the nervous and endocrine systems, controllers of body functions, operate mainly as negative feedback systems. Some of these control systems are relatively simple and easily understood; others may have several complicated inputs that balance and counterbalance each other. Both the nervous system and the endocrine system work together to control our responses to various internal and external environmental stimuli.

NEGATIVE FEEDBACK SYSTEMS

Nearly everyone who has worried about the utility bill is familiar with the thermostat. When it is set for a certain temperature and the ambient temperature drops below the set point, the electrical

FIGURE 1-1 Operation of a thermostat in the maintenance of a set temperature. When the thermostat registers a temperature below that of the set point, the relay system activates the control system. Another relay system turns on the heating unit. The heat produced by the heating unit raises the room temperature and shuts off the message to the thermostat. The response of the unit (heat) offsets the original stimulus (cold).

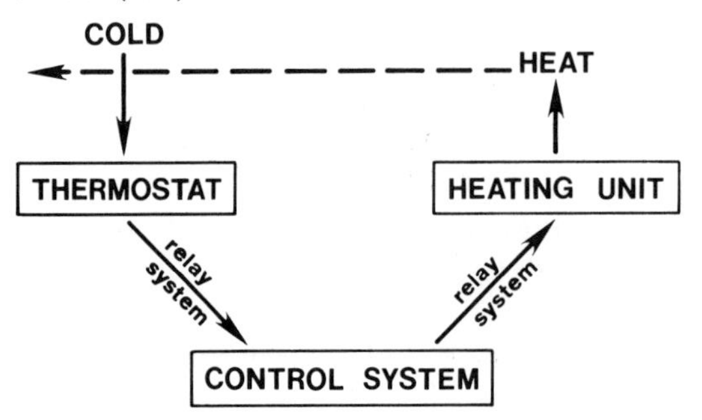

connections to the heating system stimulate the release of energy in the form of heat. When the heat causes the room temperature to rise above the set point, the thermostat shuts off the heating unit. The system, diagrammed in Figure 1-1, includes the following components:

THERMOSTAT HEAT CONTROL SYSTEM	*GENERALIZED NEGATIVE FEEDBACK SYSTEM*
the temperature of the room	*stimulus*
thermostat	*receptor*
electrical connection from thermostat to regulator of heating unit (relay system)	*afferent pathway (to integrator)*
regulator of heating unit	*integrator*
electrical connection from regulator to heating unit (relay system)	*efferent pathway (away from integrator)*
heating unit	*effector*
heat energy	*response*

The heat energy, produced by way of the pathways described above, is the response to the initial stimulus, which was the lowering of the room temperature below the set point. The response (release of heat energy) offsets the stimulus.

The operation of the thermostat is illustrative of the general way in which negative feedback systems, nervous and hormonal, control our bodies (Fig. 1-2). The aim of the stimulus-offsetting response is to maintain the internal environment of the body and of the cells that compose it. Walter Cannon, a noted American physi-

FIGURE 1-2 Components of a generalized negative feedback system. In order to keep the internal environment of the body at specific set points, the nervous and endocrine systems operate as negative feedback systems in which the response offsets the stimulus.

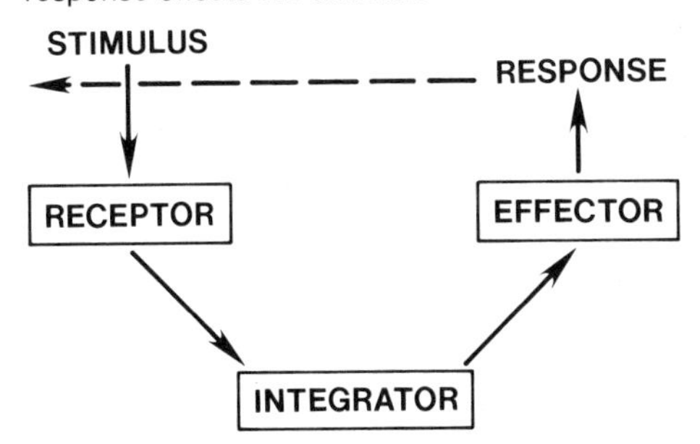

ologist, referred to this process of changing-to-stay-the-same as homeostasis (from Greek *homeo* 'same', and Latin *stasis* 'condition').

Homeostatic control systems are present in our bodies at all levels—cell, tissue, organ, and system. They regulate such functions as the endocrine control of blood sugar, arterial blood pressure, and the timing of ovulation during the ovulatory cycle. The control mechanisms of both the nervous system and the endocrine system operate to maintain homeostasis. Each will be described in terms of a negative feedback system.

NERVOUS CONTROL SYSTEMS

The components of nervous control systems are illustrated in Figure 1-3. They include:

GENERALIZED NEGATIVE FEEDBACK SYSTEM	*NERVOUS SYSTEM*
stimulus	*some form of energy*
receptor	*sense organs, nerves, etc.*
afferent pathway	*sensory root of spinal nerve*
integrator	*spinal cord, medulla, cerebral cortex, etc.*
efferent pathway	*motor root of spinal nerve*
response	*muscle contraction, gland secretion*

In the example illustrated in Figure 1-3, a reflex response removes the arm from the danger source almost before conscious comprehension of the situation occurs, and certainly much faster than the components of the response could be analyzed and/or described. The reflex response is the simplest type of nervous control mechanism. Simple reflex responses are only a small part of what we do each day. Our human activities generally involve much more complicated inputs and responses. They, too, follow definite pathways in the central nervous sytem, that is, the brain and spinal cord. Nerves from the brain and spinal cord reach nearly every structure in our bodies; they constitute the pathways along which nerve impulses travel. Sometimes these nerve impulses are consciously directed as, for example, when we are grocery shopping, feeding the baby, feeding ourselves, solving problems, or even reading this book. Unconscious activities, such as digestion, heart rate, and respiration, are also regulated through the nervous system.

Nerve impulses are carried in nerves in the form of electrochemical changes in the nerve cell membrane. These changes travel along the length of the nerve until they reach the place where the

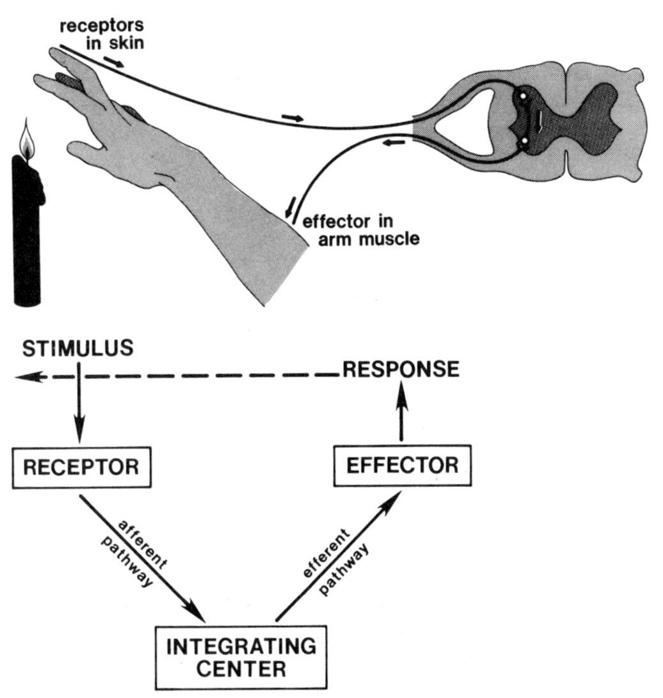

FIGURE 1-3 Some components of a nervous control system. Heat in the candle flame activates nerve endings and special heat receptors in the skin. Coded sensory message travels in a nerve to the spinal cord. Here the message is transmitted by way of the connecting nerve cell to a motor nerve, which conveys it to the skeletal muscles of the arm. Contraction of the skeletal muscles in the arm moves the hand away from the candle. Other connecting nerve cells may transmit the message to the brain.

nerve cell joins another nerve, a muscle, or a gland. There is a microscopic space at these junctions. When the space is between a nerve and a muscle it is called a neuromuscular junction. When it is between a nerve and another nerve it is called a synapse. Although the space is insignificant in size, its function is critical. It ensures that nervous impulses are transmitted one way, avoiding nervous impulse traffic jams. When nerve impulses reach the synapse or the neuromuscular junction, they stimulate the secretion of neurotransmitters. These are chemicals with a specific function. After being secreted at the end of one nerve cell, neurotransmitters diffuse across the microscopic space and attach themselves to receptors on the cell membrane of the adjacent nerve cell (or muscle or gland). Here they initiate a response. Figure 1-4 illustrates the activity of neurotransmitter substances.

Chemical compounds that act as neurotransmitters have been identified. The chief neurotransmitters are norepinephrine and acetylcholine. They are synthesized by nerve cells and are stored in vesicles near the synapse. Neurotransmitters are released by nerve

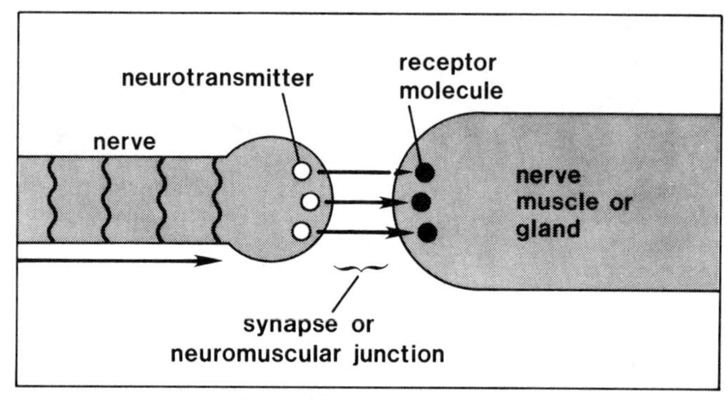

FIGURE 1-4 Neurotransmitters in nervous control systems. The nerve impulse passes down the nerve cell to the synapse or neuromuscular junction. Here the neurotransmitter substance is released. It diffuses across the synapse and is attached to receptor molecules in the adjacent nerve, muscle, or gland. The neurotransmitter-receptor complex activates another impulse in the cell that has the receptors.

stimulation and react with specific receptors. The formation and release of neurotransmitter substances and their subsequent attachment to these specific receptors are vital to the proper functioning of nervous control mechanisms. Any substance or condition that interferes with the activity of the neurotransmitter will also affect the control of specific body functions. Alcohol, a central nervous system depressant, may affect the transmission of the nerve impulse at the synapse by blocking either the synthesis of the transmitter or its attachment to the receptor. We may recall from our own experience that, at times, drinking alcoholic beverages interferes with our thinking and moving processes, particularly with our control of these processes.

Fatigue and other malfunctions in the nervous system may be related to the rise and fall in quantities of neurotransmitter substances. The measurement of neurotransmitter substances has demonstrated some relationship between our moods and the levels of neurotransmitters in our bodies; for example, mood elevation has been associated with high levels of norepinephrine, while depression is associated with low levels of this substance. The influence of chemicals on our feelings and our behavior is an active area of research in neurology, endocrinology, and neuroendocrinology.

ENDOCRINE (HORMONAL) CONTROL SYSTEMS

Whereas nervous control systems regulate relatively fast and short-range activities, endocrine control systems are related to slower, longer-range changes in our bodies. The substances that regulate these longer-lasting changes are called hormones. All the hormones, along with the glands that produce and secrete them, make up the endocrine system. The locations of the principle endocrine glands in the female are shown in Figure 1-5. Hormones have in common the following characteristics:

1. *They are effective in very small quantities.*
2. *They are synthesized by living cells which are located in specialized structures called* glands *(Fig. 1-5).*
3. *They are secreted into the bloodstream and transported by the circulatory system.*
4. *They are effective only in their target organs, where they produce a specific response.*

Hormones are physiological regulators. They speed up or slow down reactions that would proceed at different rates in their absence. For example, the hormone insulin facilitates the entrance of glucose into the cells of our bodies, where it is used as an energy source. Without insulin, glucose either stays in the blood or it is excreted in the urine.

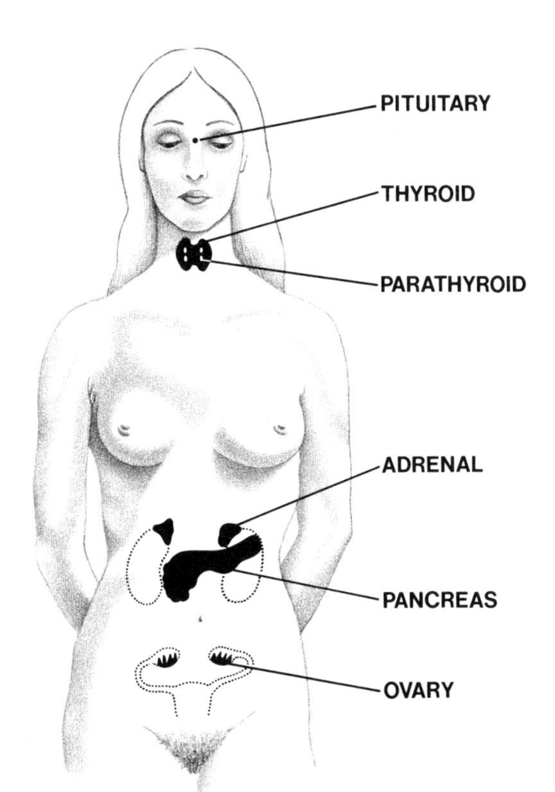

FIGURE 1-5 Locations of the principle glands in the female body.

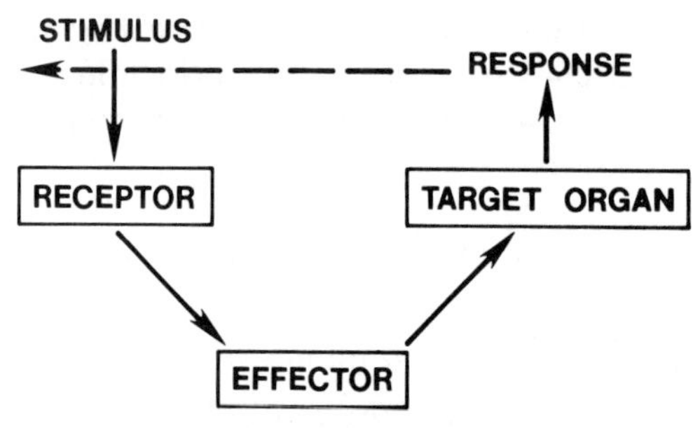

FIGURE 1-6 Endocrine (hormonal) negative feedback systems. The control of blood sugar by the hormone insulin illustrates how these systems operate. The *stimulus,* high blood sugar, affects the islets of Langerhans in the pancreas. These cells are *receptors* for the stimulus. Insulin, produced by the pancreas, acts as the *effector.* Insulin affects the *target organs,* in this case all those organs whose cells use glucose for an energy source. Insulin facilitates the entry of glucose into the cells. This *response,* glucose entry into the cell, offsets the stimulus because glucose moves from the blood into the cell.

The operation of a simple hormonal negative feedback system, such as the one involving insulin, is illustrated in Figure 1-6. In the case of insulin, the hormone is secreted by the pancreas in response to the stimulus of high blood sugar, following ingestion of food. Insulin travels in the bloodstream to all the cells of the body that have insulin receptors. The attachment of insulin to the receptors permits the entry of glucose into the cell and the lowered blood glucose "shuts off" insulin production.

Although other hormonal control systems may be more complex than the one illustrated, similar components are involved.

GENERALIZED NEGATIVE FEEDBACK SYSTEM	*HORMONAL CONTROL SYSTEM*
stimulus	*rise or fall of level of metabolic substance*
receptor	*gland*
afferent pathway	*circulatory system*
integrator	*none*
efferent pathway	*circulatory system*
effector	*release of hormone*
response	*adjustment of level of metabolic substances*

The specific response (release of the hormone) in a hormonal control system depends on the properties of the cell in the stimulated target

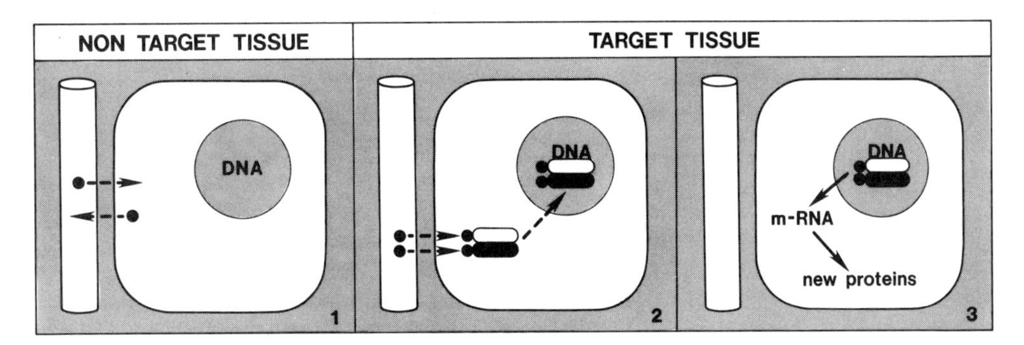

FIGURE 1-7 Steroid hormones in nontarget and target tissues. (1) In nontarget tissues steroid molecules move freely in and out of the cell. (2) In target tissues, however, hormones are sequestered by steroid receptor molecules. Each receptor binds two molecules of hormone and enters the nucleus of the cell. (3) There it becomes attached to the nuclear material, DNA. Through messenger RNA, DNA then directs the formation of new proteins. Adapted from B. W. O'Malley and W. T. Schrader, "The Receptors of Steroid Hormones," *Scientific American* 234, no. 2 (February 1976): 32–43. Copyright © 1976 by Scientific American, Inc. All rights reserved. By permission of W. H. Freeman and Company for Scientific American, Inc.

tissue. Hormones, as we know, are transported to all the cells of the body by means of the circulatory system. Why then, we may ask, do hormones produce their effects only in some parts of the body and not in others?

Hormones act on target tissues because there are specific receptors for them in the cells of the target tissues. Not all the tissues in our bodies have the same specific hormone receptors.

The interaction between the hormone-receptor combination and the function of the cell depends on the chemical structure of the hormone. Steroid hormones, such as estrogen and progesterone, affect the operation of their target cells in a way different from that of nonsteroid hormones. We need to understand both types of interactions so that we can analyze the effects of other substances (some of which may be hormones) on the response of the hormone-stimulated cell.

Steroid hormones, carried in the circulatory system, reach all the cells, but affect only certain ones. Figure 1-7 illustrates this selective process. In nontarget tissues, steroid hormones diffuse[1] from the capillary to the space between the cells and in and out of the cell (1). In target cells, on the other hand, steroid hormones are *bound to receptors* when they reach the cytoplasm[2] of the target

[1]Diffusion is a process whereby molecules move from one place in the body to another. Movement follows a concentration gradient; that is, the molecules move from the places where they are more concentrated to where they are less concentrated.

[2]Cytoplasm is the part of the cell within the cell membrane and outside of the nucleus.

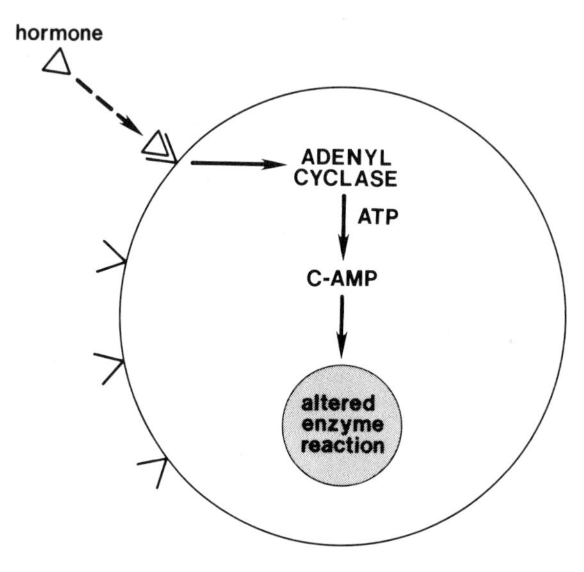

FIGURE 1-8 Activity of nonsteroid hormones in target tissues. The hormone attaches to a receptor on the cell membrane and forms a hormone-receptor complex. This activates an enzyme, adenyl cyclase, which produces cyclic AMP from ATP. The cyclic AMP acts as a "second messenger" to bring about an altered enzyme reaction, which is one effect attributed to hormone activity.

cell (2). The hormone-receptor combination then fastens itself to the chromosomes[3] in the cell nucleus (3), where it influences the programming of the cell in order to produce protein. New proteins may be synthesized in the hormone stimulated cell.

The interaction between the nonsteroid hormone and the cell it stimulates is illustrated in Figure 1-8. The hormone attaches to receptors on the cell membrane. As a result of the hormone-receptor interaction, an enzyme within the cell, adenyl cyclase, is activated. The enzyme in turn causes a transformation of a substance within the cell cytoplasm, adenosine triphosphate (ATP), to a product called cyclic adenosine monophosphate (c-AMP). The cyclic AMP then directs the reaction, which is usually attributed to the hormone.

Hormones that regulate the function of the ovary are peptide (nonsteroid) hormones. During a woman's reproductive life, her ovaries respond to these peptide hormones. However, as the ovary ages, it becomes less responsive to hormonal stimulation. We can see from Figure 1-8 that nonresponsiveness such as that which characterizes the aging ovary can occur at several different biochemical sites in the pathway between the attachment of the hormone to the receptor and the specific response of the ovarian cell.

Hormones, then, act on specific target tissues of our bodies because the cells that compose these tissues have receptors for them. The products of the hormone-stimulated cell regulate body functions through negative feedback systems which may be simple or complex. Because hormones are effective in such small amounts, we need

[3]Chromosomes (see Chapter 2) contain DNA, the genetic material that controls heredity.

to pay particular attention to the effects of ingested hormones (such as birth control pills and estrogen replacements) on homeostatic control systems that regulate all of our metabolic[4] functions.

NEUROENDOCRINE CONTROL SYSTEMS

The interrelatedness of the two systems that control our bodies is the subject of its own branch of biology and medicine, neuroendocrinology. For our purposes we do not need to understand the whole science, but we should be familiar with the mechanisms by which nerves and hormones can influence each other. We may experience the results of neuroendocrine interactions as postpartum depression, as boarding school amenorrhea,[5] or as difficulty in breast-feeding our babies.

The influence of the *endocrine system on the nervous system* can be inferred if we remember that (1) the blood, which contains the hormones, circulates to all parts of the body, including the nervous system; and that (2) cells in the nervous system may have receptors for hormone molecules. In order to understand how the *nervous system influences the endocrine system,* we need to visualize the anatomy of the brain in general, and of a specific area of it, the hypothalamus, in particular. Figure 1-9 illustrates the location of the hypothalamus, the connection between the brain and the pituitary gland. In the hypothalamus, specialized cells, called neurosecretory cells, can translate a message from the nervous sytem into a chemical message which is carried in the circulatory system. In special centers of the hypothalamus, some neurosecretory cells produce hormones (oxytocin and vasopressin) that are carried to the posterior portion of the pituitary gland. They are stored there and released following appropriate nervous stimulation. For example, the sucking of the baby's mouth during nursing stimulates the release of oxytocin, which functions to release milk from the nipples.

Other neurosecretory cells in the hypothalamus produce releasing factors, which, in turn, control the release of other hormones from the anterior part of the pituitary gland. Figure 1–9 shows the anatomical connections between the nervous and hormonal control systems. The nerve cells at (A) and (B) in the hypothalamus produce oxytocin and vasopressin; those at (C) produce the releasing factors.

We have seen that mechanisms exist in the hypothalamus for the translation of nerve impulses into chemical messages, and thus, for the nervous system to influence the endocrine system. An example

[4]Metabolism is the sum of all the physical and chemical processes that maintain life. It also refers specifically to transformations that furnish energy to the body.

[5]Amenorrhea is the absence of menstruation.

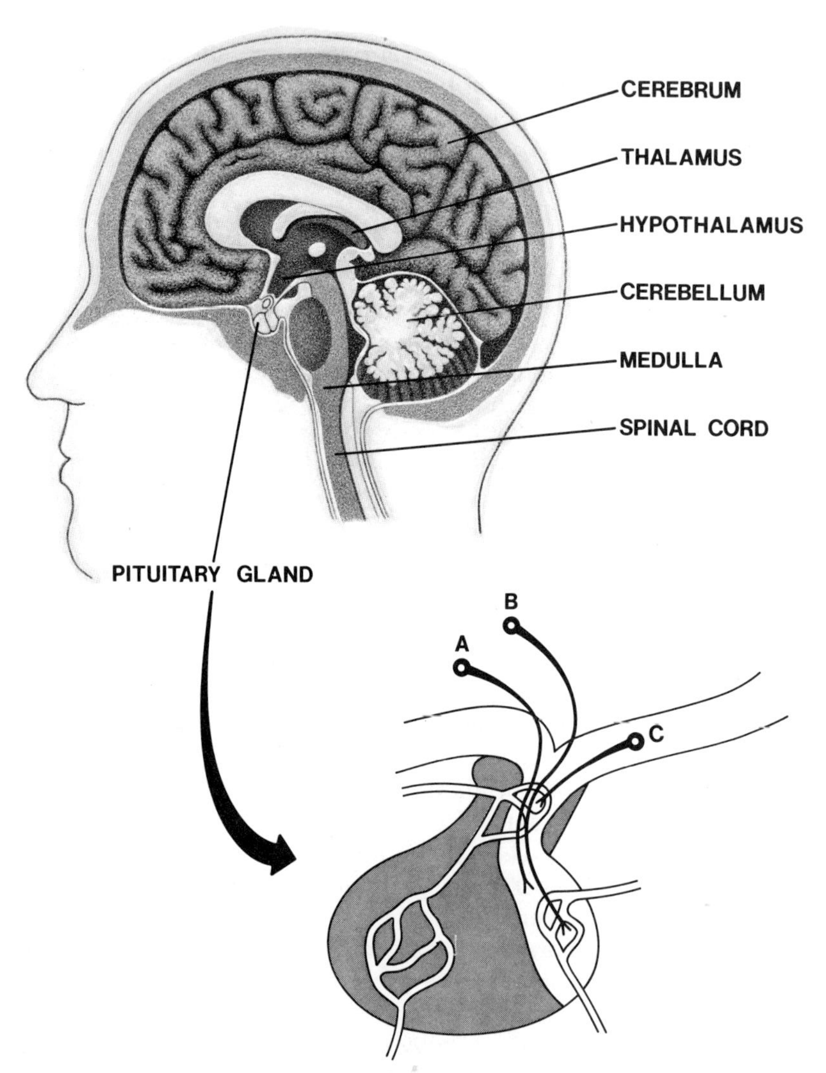

FIGURE 1-9 Anatomical relationship between the nervous and endocrine systems.

of such influence may be found in "boarding school amenorrhea." This is a condition in which young women away from home for the first time, usually to attend boarding school or college, may experience a temporary absence of menstrual bleeding unrelated to pregnancy exposure. It has been proposed that the stress caused by encountering new surroundings, new people, strange food, and strange schedules may cause inputs from the central nervous system that interfere with the hypothalamic releasing factors, and, thus, via the pituitary gonadotropins, with ovarian function. Sometimes anxiety about being pregnant can delay menstruation by the same neuroendocrine interaction.

Many other examples of neuroendocrine regulation of our bodies and their functions could be cited; subsequent chapters will do just that. If we understand, now, in a general way, how these systems cooperate we can begin to appreciate the *balance* within each system and between the two systems.

MEASUREMENT OF HORMONES AND NEUROTRANSMITTERS

The hormones that define and determine our femaleness are secreted in different quantities at different stages in our female cycles. These hormonal changes have been measured with greater and greater accuracy as new methods were perfected in the laboratory. For example, the hormones that regulate the ovulatory cycle, follicle-stimulating hormone (FSH) and luteinizing hormone (LH), are secreted in different amounts during different phases of a woman's life. Figure 1-10 illustrates gonadotropin secretion in women during prepubertal, reproductive, and postreproductive life. During early puberty these hormones are produced in very small amounts. Production increases during late puberty, and becomes cyclical during reproductive life. As the negative feedback effects from the ovary cease during the menopause, gonadotrophic hormones are produced at sustained higher levels. Being able to measure gonadotropins and gonadotropin-like substances accurately has helped to define:

1. *early vs. late puberty*
2. *stages of the ovulatory cycle*
3. *very early stages of pregnancy*
4. *stages of the menopause.*

FIGURE 1-10 Life cycle changes in gonadotropin production in women. From S. C. Yen et al., "Causal Relationship between the Hormonal Variables in the Menstrual Cycle," in *Biorhythms and Human Reproduction,* ed. M. Ferin (New York: John Wiley & Sons Medical Division, 1974), p. 220. By permission of John Wiley & Sons, Inc.

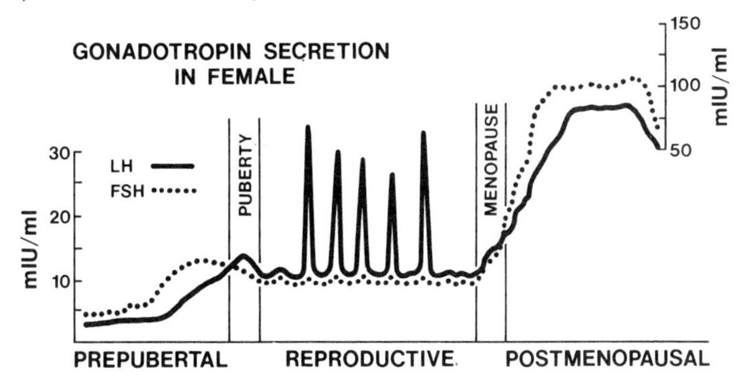

Good methods for measuring these and other hormones have greatly facilitated diagnosis of diseases as well as of reproductive status. Reliable methods for measuring neurotransmitters are also available. Therefore it would seem that quantitive measurement of hormones and/or of neurotransmitters should be an integral part of any study that attempts to demonstrate a correlation, in the human female, between hormones and behavior or between genes and hormones.

The next chapter will outline the genetic determinants of femaleness and the relationship between the chromosomes we inherit and the sexuality we develop. There we shall discover that hormone production begins early in fetal life.

2 Chromosomes and Genes

We were programmed before we were born to develop as females. What does it mean to be female? Genetically, it means that we have inherited one X chromosome (and 22 other chromosomes) from each parent. The relationship between the inherited chromosomes and our development as females is the subject of this chapter.

OUR GENETIC INHERITANCE

The DNA Story

Chromosomes, genes, and DNA—some of us know the relationship among them, some of us do not. We all know, in a general way, that children are like their parents, only different. The specific reasons for these similarities and differences are in the chromosomes, the genes, and the DNA.

The DNA story began in the mid-1800s in an Austrian monastery. In the garden of this monastery, Father Gregor Mendel cross-pollinated and counted pea plants. He observed that certain "characters" of the parents were seen in the offspring in constant ratios. His work was rediscovered in the early 1900s when Thomas Hunt Morgan, an American geneticist, greatly expanded our insight into the mechanisms underlying the inheritance of Mendel's "characters." Morgan knew that the cell nucleus contained structures that were selectively stained by certain dyes. The structures colored by the dyes were named chromosomes (from Greek *chromo* 'color', and *soma* 'body'). Morgan and his students were soon able to show that the genes, located on the chromosomes, determined the characteristics of plants, fruit flies, and even people. Further progress in understanding the mechanisms of heredity was made when two American scientists, George Beadle and Edward Tatum, showed what the genes *did.* They were able to show how genes operate to produce proteins.

So far the scene was set and the preliminary descriptions were in, but where was and who was the leading character? A material isolated from such diverse sources as salmon sperm, human pus cells, and bacteria acted like genes. Its chemical components—phosphates, sugars, and bases—were identified, and the genetic material itself was called *d*eoxyribo*n*ucleic *a*cid (DNA).

Many dedicated scientists worked to solve the structure of DNA. Many contributed one element to understanding what the molecule looked like. The synthesis of these elements, which yielded the precise structure of DNA, was the work of an American, James Watson, and an Englishman, Francis Crick.

Figure 2-1 illustrates the relationship between the chromosomes and the DNA, which is their principle component. At the bottom of the figure, the double helix configuration of DNA has been magnified to show the backbone of the helices and the base pairs that join them. The backbone of the double helix is made up of sugars and phosphates. The base pairs make up the central part of the helix. The bases are abbreviated A (adenine), T (thymine), G (guanine), and C (cytosine). Adenine always pairs with thymine, and guanine always pairs with cytosine. The sequence of these bases and their complementary pairing is a key to understanding how chromosomes function to duplicate themselves and to synthesize proteins. Our genes, segments of the DNA molecule, direct the orderly sequence of events known as human development.

Knowing about Our Own Genes: Family Histories and Genetic Counselling

At the moment of our conception, the 23 chromosomes in our mother's ovum (egg cell) were united with the 23 chromosomes in our father's sperm cell, creating a fertilized ovum with 46 chromosomes. When this cell divided, the chromosomes duplicated themselves so that each new cell (except for our own ova) would also contain 46 chromosomes. As further cell division took place, different chromosomes were "turned on" so that our cells became differentiated into specialized cells of various types, including blood cells, bone cells, liver cells, muscle cells, nerve cells, and skin cells. Thus the chromosomes we inherited from our parents determined the range of our total physical development, including our sex. The chromosomes of genetic females and genetic males are illustrated in Figure 2-2.

Some of these chromosomes may contain faulty information

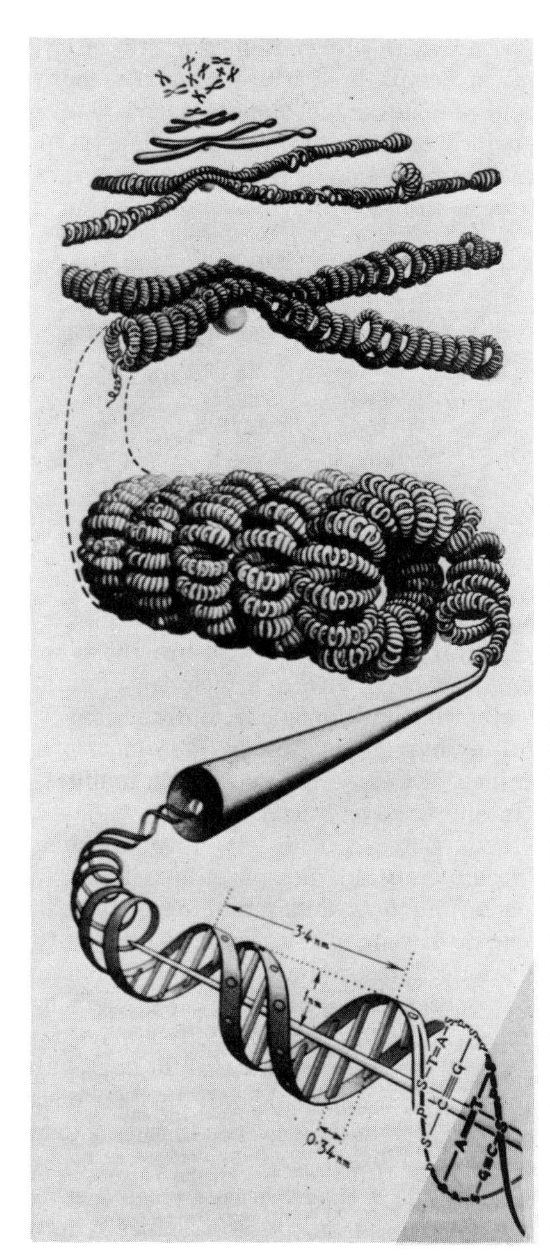

FIGURE 2-1 Chromosomes and DNA. Each chromosome is made up of two chromatids united at the centromere. A chromatid is a protein-coated strand of multicoiled DNA. The DNA within the protein coat consists of the bases adenine (A), thymine (T), cytosine (C) and guanine (G). These bases are united through a sugar, deoxyribose (S), to phosphate groups (P). The double and triple hydrogen bonds between the bases are broken prior to cell division when DNA is producing more DNA or before protein synthesis when, through RNA, DNA directs the formation of new proteins in the cell. Diagram from Roger Warwick and Peter L. Williams, *Gray's Anatomy,* 35th British ed. (London: Longmans Co. Ltd., 1973), p. 15. With permission of Churchill Livingstone.

in the form of changes in the DNA. A change in the sequence of bases in DNA, the deletion of a base, or the miscopying of a base during chromosome replication can all lead to faulty transmission of information. A defective gene may result in mistakes in the proteins in our bodies or, in some cases, in no proteins. As parents, we can pass these defects on to our children.

Defective genes or other chromosomal abnormalities, such as

those that occur when chromosomes break, often result in early developmental problems in fertilized ova. As many as 50 percent of early conceptuses never develop to term. They are aborted spontaneously before the eleventh week of gestation, indeed, perhaps even before a woman knows she is pregnant. The fate of any 1,000 fertilized ova, assuming no medical intervention, has been estimated as follows:

	Number	*Percent*
Lost before implantation	250	25
Lost before pregnancy diagnosed	150	15
Spontaneous abortion	100	10
Normal infants	400	40
Stillborn infants	10	1
Special care infants	90	9

We may note the category of special care infants, which comprises 9 percent of total conceptions. Who are these special care infants? They are children born with neurological, immunological, muscular, metabolic, or other physiological malfunctions that may interfere with the full enjoyment of human life. Some of these birth defects "run in families." Do they run in our own families?

If we intend to have children of our own some day in the future, we should try to find out if we have any inherited diseases in either family. We can question our parents and our parents-in-law and we can broaden our own understanding of inheritance patterns. We are fortunate to live in the age of prenatal chromosome diagnosis and other prenatal examinations. Skilled genetic research can let us know percentage possibilities of conceiving a child with an inherited and/or inheritable disease. In earlier times the only way a doctor or the prospective parents could deal with a genetic disease was to await the birth of a child who had the disease. Modern medicine has given us alternatives, as the following example illustrates.

Mary and Joe were 23 and 24 years old respectively when they married. Both were perfectly well themselves. Mary's brother, however, had died at the age of 15 from muscular dystrophy. At the time of her brother's death, the parents were told that Mary was indeed a possible carrier of muscular dystrophy and was therefore at risk for having a male offspring with this disease. No effort was made to actually determine if she was a carrier, and as the years rolled by and Mary's marriage approached, no one suggested genetic counselling again. The very first pregnancy yielded a beautiful boy who appeared entirely normal. However, when his son was 3½ years old, Joe noticed that

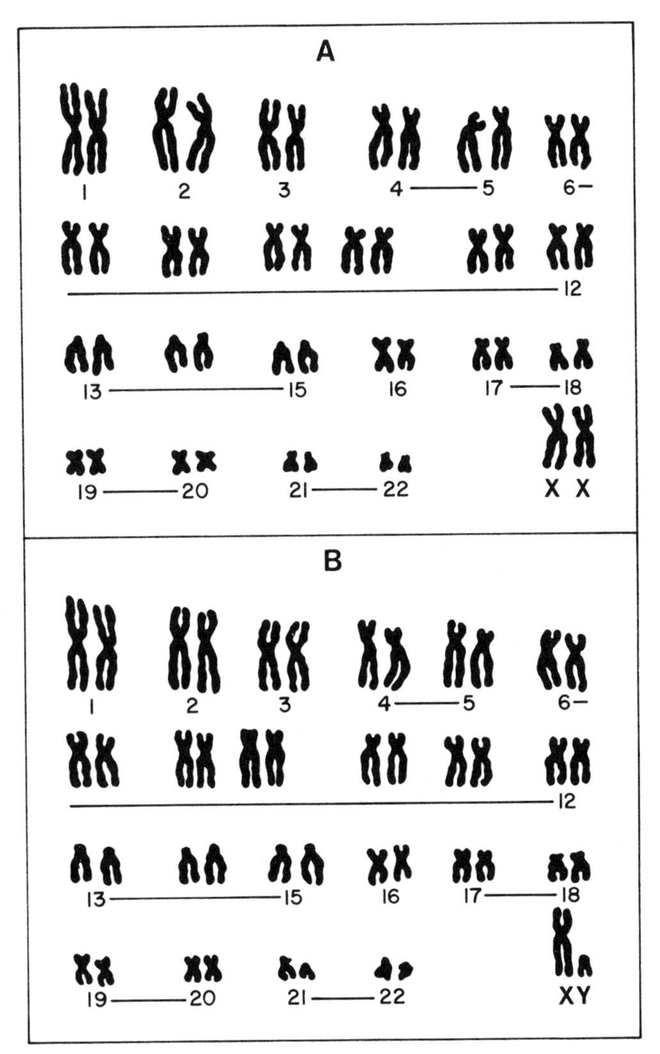

FIGURE 2-2 The human chromosomes. The set of chromosomes designated A are female; those designated B are male.

he had difficulty climbing the stairs and even getting up from a sitting position on the floor. The diagnosis of muscular dystrophy was made in the same week that Mary's second pregnancy was confirmed.

The attending general practitioner counselled the parents that Mary indeed was a carrier (confirmed by a blood test) and that in the future she would have a 50 percent risk that any male offspring would be affected by muscular dystrophy. The physician advised, however, that in the preceding year it had become possible to predict fetal sex early enough in pregnancy to offer the parents elective abor-

tion if a male fetus was indeed found. Mary and Joe opted for the amniocentesis that showed that the fetus was a male, and they elected to terminate the pregnancy. Using prenatal diagnosis studies, they subsequently had two normal girls.[1]

The amniocentesis procedure mentioned in the preceding narrative may have been explained to some of us before. We may already know what it is, who should have it done, and what it can tell us about our unborn children. We may even have friends whose anxieties about inherited diseases were set to rest for the remainder of their pregnancy, following an amniocentesis. We may also know women who elected to have an abortion when amniocentesis revealed a fetus with Down's syndrome (mongolism).

At the present time amniocentesis is usually performed only when there is an adequate reason for it. Pregnancy after the age of 35 is one such reason, because the risk of Down's syndrome increases along with the mother's age. Down's syndrome occurs once in about 2,300 live births when the mother is under 20. In women over 40, this syndrome may occur once in 100 live births. Over age 45, the risk increases to 1 in 45 live births. Women like Mary, who know that they might conceive children with genetic defects, should request amniocentesis.

An amniocentesis is a simple outpatient procedure for the high-risk woman. It is usually done between the fifteenth and seventeenth week of pregnancy, when the total volume of amniotic fluid is more than 100 cc. After the overlying skin of the abdomen is anesthetized, a needle within a needle is inserted through the abdomen and uterine wall into the amniotic fluid surrounding the fetus. Some of the fetal cells, which were sloughed off from the skin, respiratory, or urinary tract, are drawn off with the fluid and prepared for analysis. The fetal cells that are taken from the fluid are grown in the laboratory in a special nutrient medium. This process is called culturing. Then the chromosomes are analyzed. Biochemical tests may also be carried out on the amniotic fluid. Both types of analyses require approximately three weeks. Figure 2-3 illustrates the procurement of amniotic fluid and culturing of cells. The procedure itself is demonstrably safe and poses no danger to the fetus when it is done by a skilled physician.

[1]Aubrey Milunsky, M.D., *Know Your Genes* (Boston: Houghton Mifflin Company, 1977), p. 8. Copyright © 1977 by Aubrey Milunsky, M.D. Reprinted with permission. This extremely interesting study of human genetics and genetic counselling is written by a recognized authority in the field. The reader who suspects she has a genetic disease in her family, or who is interested in knowing more about human heredity is advised to read this book as a starting point.

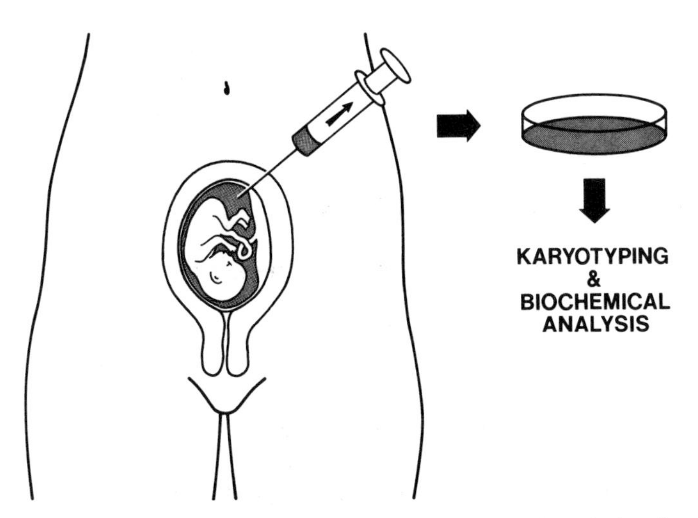

FIGURE 2-3 Amniocentesis. The needle is inserted through the abdominal muscles and uterus into the amniotic cavity. Cells and fluid are withdrawn. Culture of cells in the fluid permits chromosome analysis. The fluid itself can be studied for the presence of abnormal enzymes.

The main purpose of prenatal testing, which now includes procedures other than amniocentesis, is to detect birth defects early enough for the pregnancy to be terminated if this is the wish of the parents. If the test should indicate that the fetus is abnormal the decision to abort or to carry the pregnancy to full term may be physically, morally, and emotionally difficult for the parents to make. Parents who decide not to terminate a pregnancy involving a fetus with a known birth defect can prepare themselves beforehand for the care of the child. On the other hand, approximately 95 percent of prenatal testing shows that none of the genetic abnormalities being tested for are present, so that high-risk pregnancies in these cases can proceed without undue anxiety.

A great many families are fearful of having additional children, or desperate to terminate an unplanned pregnancy because of recurrent risks of Down's syndrome, Tay-Sachs disease, and other genetic disorders. Amniocentesis and other appropriate tests can help these families plan further children and to complete the vast majority of pregnancies because of the assurances that the baby will not be affected with the disease they fear most. In a small number of cases, less than 5 percent, the couple may have to choose between having an affected child and terminating the pregnancy, admittedly a wrenching decision for most couples. [Omenn, 1978]

Knowing about our own genes may be like knowing about anything else: We may not like what we learn. Accurate knowledge, however, is always a better basis for decision than fear or superstition. We can make a better choice, in accordance with our own moral values, when we know beforehand just what the choice involves.

OUR DEVELOPMENT AS GENETIC FEMALES

The Influence of Sex Chromosomes

The genetic determination of our femaleness occurs at the time of our conception. Our mothers contribute 22 regular chromosomes (autosomes) plus an X chromosome, while our fathers contribute 22 autosomes and *either* an X or Y chromosome. As Figure 2-4 shows us, it is the male parent who actually determines the sex of his child. If he is familiar with gambling, he will quickly see that five preceding female births will not change the odds for the conception of a male child next time.

Male and female embryos develop in a similar fashion until about the seventh week of pregnancy. Their chromosomes are actively directing the synthesis of cells that will differentiate to form the body parts—such as muscles, nerves, heart, and blood vessels—that are the same in both sexes. After the seventh week, a small part of the male embryo's Y chromosome begins to make a special protein. This protein, called the H-Y antigen, is made through the mechanisms that direct protein syntheses in cells. Figure 2-5 illustrates the relationship between the DNA (genes) on the chromosomes and specific proteins, such as the H-Y antigen. If other DNA-directed mechanisms are functioning adequately, the H-Y antigen directs the following changes:

1. *The gonads, undifferentiated sex organs at this time, are directed toward the formation of testes.*

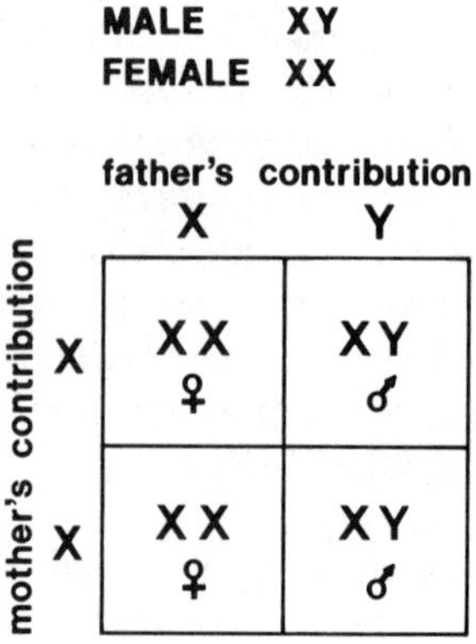

MALE XY
FEMALE XX

father's contribution

mother's contribution	X	Y
X	XX ♀	XY ♂
X	XX ♀	XY ♂

FIGURE 2-4 Probability of male or female conception.

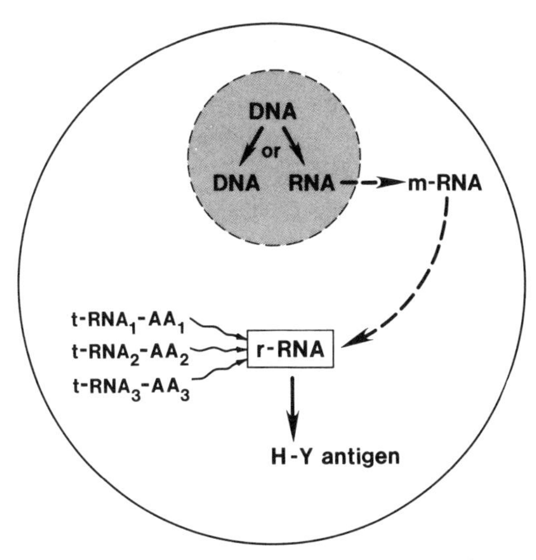

FIGURE 2-5 DNA and protein synthesis. Chromosomal DNA can produce either more DNA (if the cell is dividing) or RNA (if the cell is synthesizing protein). In the genetic determination of gonadal development, nuclear DNA produces RNA which passes through nuclear pores to the cytoplasm. This messenger RNA (m-RNA) moves to the ribosomes where it functions as a template for protein synthesis. Ribosomal RNA consists of specific base sequences. Smaller units of transfer RNA (t-RNA) attach themselves to amino acids (AA) in the cytoplasm of the cell. These amino acids, when deposited sequentially on the ribosomal RNA ($AA_1 + AA_2 + AA_3$, etc.) will form specific proteins such as the H-Y antigen.

2. *The newly formed testes begin to produce two other types of molecules:*
   a. a protein, MRF (Mullerian-repressing factor), which inhibits development of female internal and external reproductive structures
   b. testosterone, which stimulates development of male internal and external reproductive structures.

Thus, the basic embryonic plan of human development is female. In the absence of MRF and testosterone, the internal and external genital structures develop as female, and the embryonic ovaries produce the female hormones, estrogen and progesterone. Figure 2-6 illustrates sexual development in the presence and absence of the Y chromosome. The presence of the Y chromosome in the male stimulates synthesis of the H-Y antigen. The embryonic gonads develop into testes, which produce the substances necessary for the development of male reproductive structures. Basically, we are females because we lack the Y chromosome.

The X chromosome is more than just a sex-determining chromosome. Other genetic characteristics, including some human diseases, are carried on the X chromosome. These diseases include hemophilia, colorblindness, muscular dystrophy, and a type of eye disease. When a male has an X chromosome that has these traits, the Y chromosome does not exert an opposing effect on protein synthesis and thus he shows the trait. Because females have two X chromosomes, the probability that sex-linked recessive traits will be expressed is diminished. Although they themselves do not show the disease, women may be carriers. The disease in question will be seen more often in their sons than in their daughters (as we saw above in the case of Mary and Joe).

XX IN OUR CELLS Those who study cells (cytologists) can usually tell the difference between XX and XY individuals without seeing the whole individual. They can tell by the cells alone. In females, about one-fourth of the cells in any given tissue will have small,

FIGURE 2-6 Chromosomes and sex determination. The Y gene directs the synthesis of the H-Y antigen (1). Under the stimulus of the H-Y antigen the neuter gonads develop into testes (2). The synthesis of MRF (3) and testosterone (4) by the fetal testes promotes development of male external genitalia and secondary sex characteristics. Gene-directed metabolic disorders may interrupt any of these four processes and result in ambiguous sex determination.

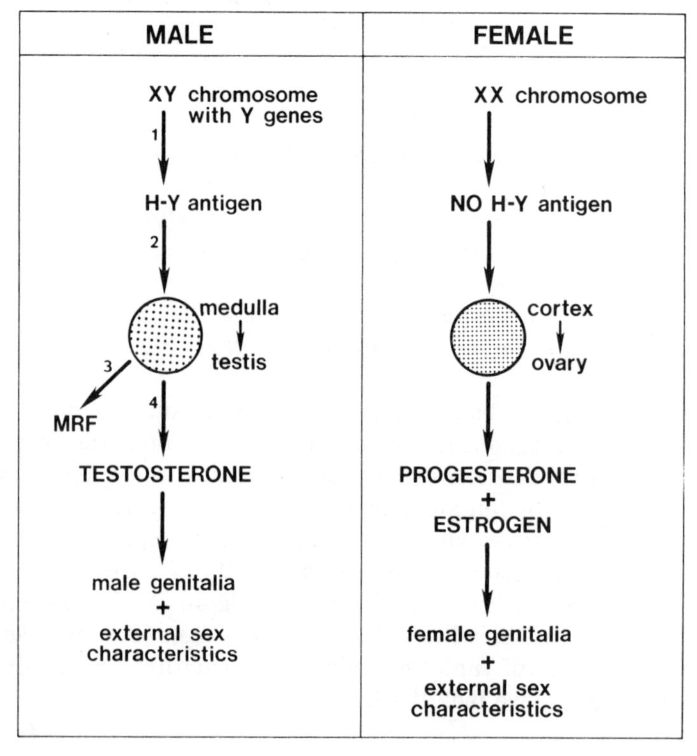

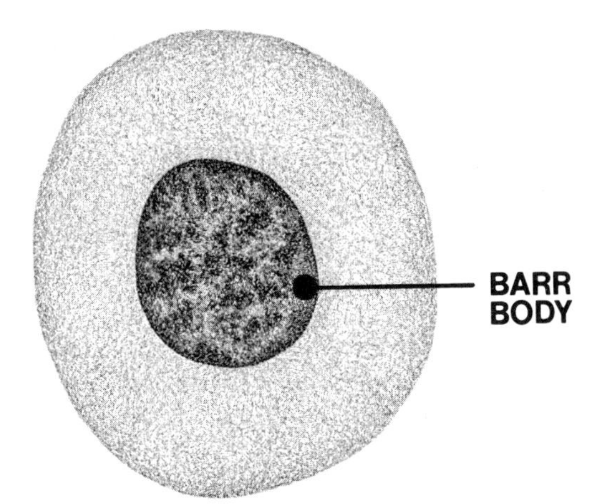

FIGURE 2-7 A female cell with Barr body.

darkly staining bodies (called Barr bodies) just inside the nucleus. This characteristic female cell is illustrated in Figure 2-7. The differences between sexes can also be demonstrated by studying certain white blood cells. Why should anyone want to do this? At least two different occasions may warrant the verification of genetic sex: (1) Some cases of hormonal imbalance may result in ambiguous external sexual organs. If corrective surgery is contemplated, it can be directed toward the complementation of the individual's genetic sex. (2) In athletic competition, particularly track and field events, cytological screening tests are often used to prevent genetic males from competing in women's events.

In Klinefelter's syndrome, which occurs in about 0.1 percent of male births, individuals with a Y chromosome have Barr bodies in some of their cells. They are genetically XXY, and possess male internal and external reproductive structures, with testes in which sperm are not produced normally. Genetic females with Turner's syndrome have no sex chromatin. The individual with this genetic disorder has only one X chromosome, and has physical abnormalities that include rudimentary nonfunctioning ovaries.

THE RESEMBLANCE OF XY AND XX

When Alice was born, her mother noticed that she had an enlarged clitoris, as well as vulval structures which did not seem to be well developed. At the age of 10, Alice was hospitalized for a leg injury, and her abnormal sex structures were noticed. The attending physician recommended to the mother that she seek further advice.

Six months later the girl was examined by a specialist in endocrinology, who observed a "clitoris" about an inch in length with a "nodule" on the right labia majora. No vaginal opening was described. Sex chromatin studies were negative. A working diagnosis of testicular feminization was made, and it was recommended that the female gender role should be continued. When Alice was about 12, a laparatomy was performed. At the time of the operation no uterus, oviducts, or ovaries were seen. Two cords were observed which terminated in masses located in the labia majora. The surgeon assumed these masses to be testes, and he was later proven to be correct.

Immediately after the operation, the surgeon walked into the waiting room and announced to the parents that Alice, their child, was really a boy. The parents were completely unable to deal with the consequences of this announcement and the stir it would cause in their small rural town. A pediatrician took over the case, and recommended that the prior gender role be reinforced. However, it took years for the family to recover from the aftereffects of one surgeon, apparently unfamiliar with the full implications of a change in gender role assignment.

Later a gynecologist removed the gonads, which were described as immature testes. A vaginal opening was constructed surgically and the patient, at 18 showed normal female development, including breast development, pubic hair and external genital structures.[2]

Testicular feminization is a rare congenital birth defect. Alice was born with a Y chromosome, which had produced H-Y antigen. However, the undifferentiated gonad, in her case, did not respond to testosterone. It is possible that receptors (see Chapter 1) for testosterone are lacking or inadequate in some individuals. It is also possible that some individuals lack enzymes necessary to metabolize testosterone into its active intermediates. Female internal structures did not develop in Alice because MRF was secreted; however, she appeared to be female because estrogen was secreted by the rudimentary testes.

Other genetic-hormonal disorders may cause problems in the identification of sex in the individual. Male genital development may occur in genetic females who are exposed to testosterone or other androgens from an external source during the eighth to thirteenth week of pregnancy. Such individuals are called pseudohermaphrodites. A pseudohermaphrodite is an individual, like Alice, with the genetic constitution of one sex and the external genitalia of another.

Tumors of the adrenal gland later in life also produce testosterone, which can lead to a virilization of the genetic female. These individuals have normal female internal and external sexual organs, but have some male external characteristics such as facial hair.

Obviously, skilled counselling is required for persons who have acted as males or females all their lives, and who are suddenly

[2]Lytt I. Gardner, "Genetic Counselling," in *Endocrine and Genetic Diseases of Childhood* (Philadelphia: W. B. Saunders Company Publishers, 1969) pp. 1030–31. Reprinted with permission.

confronted with evidence that they were originally programmed to develop otherwise. Genetic sex can be side-tracked in the physical expression of male or female sex structures because of a hormone imbalance of some kind. Can the expression of homosexual and heterosexual behavior also be explained in terms of gene-related hormonal mechanisms?

GENETICS AND SEXUAL ORIENTATION

The genetic basis of sex in human beings has been described. The Y chromosome is male-determining. In the absence of the Y chromosomes, we develop female characteristics, which include:

1. *internal structures—ovaries, oviducts, uterus, vagina*
2. *external structures—vulva (clitoris, labia majora, labia minora)*
3. *hormone production—estrogen and progesterone*
4. *external female characteristics—breast development, voice, deposition of body fat.*

Genetic sex only programs certain parts of our sexual behavior, and we know in this case that the whole is definitely more than the sum of the parts. Our sexual behavior, including our assumption of female roles, is profoundly influenced by complexly interwoven cultural, sexual, and moral considerations, as well as by nervous and hormonal control systems. Gender identity refers to an individual's self-image; it results from the factors illustrated in Figure 2-8.

If the expression of genetic sex may sometimes be waylaid by hormonal influences, what about the relationship between genes, hormones, and sexual behavior? Specifically, is homosexual behavior the result of some genetic or hormonal imbalance? Is there a genetic/biologic basis for homosexuality?

Male and female subjects in studies of homosexuals and their hormones have been genetic XX females and XY males. We know that the Y chromosome directs the formation of male internal and external structures. Subsequent to the differentiation of the testes and the production of testosterone, it is possible that hormones influence the central nervous system, and then program patterns of hormone secretion and sexual behavior.

Before we examine the evidence, we should realize that, in the human, the relationships between hormones and behavior are far from clear. Definitions of "male" and "female" behavior are so culturally and socially dependent that the role of the hormones is hardly ever considered, much less measured. In animals, however, these behavior patterns are well defined; they include copulatory posture

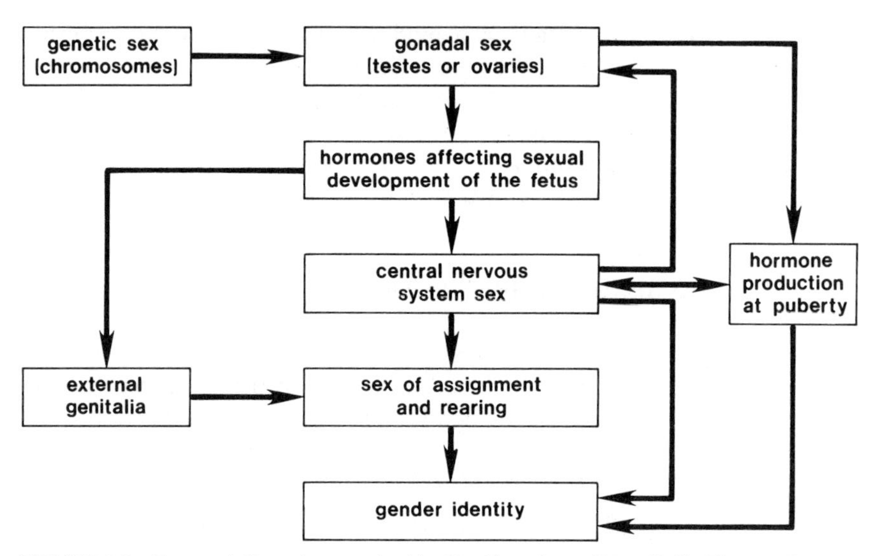

FIGURE 2-8 Factors influencing gender identity. From Leon Speroff, "Is There a Biologic Basis for Homosexuality?" *Contemporary Ob/Gyn,* Medical Economics Co. 12 (August 1978): 65–74. By permission of McGraw-Hill Book Co. and Leon Speroff, M/D.

in laboratory rats and mice, urinary posture in monkeys and dogs, and juvenile play patterns in monkeys. Animal studies show that there is a critical period during development when hormones, particularly estrogen, interact with receptors in the brain. Groups of animals can be experimentally treated with hormones during uterine and perinatal life. Results obtained from large numbers of animals can be studied in terms of amounts and kinds of hormones administered and effects on precise forms of sex-related behavior.

What we do know about the effects of hormones on human sexual behavior comes mainly from what are termed "experiments of nature." Some biochemical abnormalities may cause human fetuses to be exposed to unusual hormone influences. Three types of these experiments can be cited:

1. *Some women have taken steroid hormones early in pregnancy to prevent spontaneous abortions. Although this medication had a significant effect on the incidence of vaginal cancer in their daughters (see Chapter 8), no effect on their sex-specific behavior was noted.*
2. *A genetically determined enzyme defect in the adrenal gland causes some fetuses to be exposed to large amounts of male hormones. After birth, the administration of female hormones can correct the external manifestations of androgen effects in the genetic females. Has there been any effect on behavior because of hormonal influences during what could have been a critical period in sexual development? Yes and no. Girls with this syn-*

drome[3] *are tomboys and spend little time in expected maternal roles such as playing with dolls. They seem to have normal ovulatory menstrual cycles and they consider themselves female. Boys with this syndrome are no different from other boys in their behavior.*

3. *Some genetic males are pseudohermaphrodites because of abnormal inherited development pattern. These individuals cannot metabolize testosterone into the substance that is responsible for the expression of male external genital structures, and so they are reared as girls until puberty. At that time testosterone metabolism is normalized; the testosterone affects the descent of the testes, penis formation, voice pitch, and muscle development. A seeming transformation form "girls" to men occurs, and the men have no difficulty in assuming male gender roles (Imperato-McGinley et al., 1977).*

Studies of hormone secretion in male homosexuals and heterosexuals have shown that there is no significant difference between the two groups in the secretion of testosterone. At the same time, no significant elevation of estrogens was demonstrated (Friedman et al., 1977). It appears that the Y chromosome-directed processes in the homosexual male operate normally as far as hormone production (processes 3 and 4 in Figure 2-6). Similarly, plasma hormone levels in female homosexuals have been compared to those of matched heterosexual controls. Plasma testosterone was found to be 38 percent higher in the homosexual group (Gartrell et al., 1977). Another study, which measured urinary metabolites of estrogen, progesterone, and testosterone, found no significant differences between the two groups of women (Griffiths et al., 1974). Both studies of women were based on single samples of plasma and urine. Although these samples were collected on approximately the same day of the menstrual cycle for all subjects, it might be well to consider analyses of consecutive daily specimens from the same subjects with each woman serving as her own control. Cyclical variations in hormones could then be compared in the two groups, and some conclusions about differences or similarities in these women might be reached.

At the present time it is difficult to assay the role of hormones in homosexual behavior, as well as in other manifestations of human behavior. "It is likely that each individual has a neurohormonally determined predisposition to a particular sexual development which will be expressed if the right social environment is encountered" (Speroff, 1978). Further studies of the relationship between hormones and human behavior may indicate that it is the social environment

[3]Congenital adrenal hyperplasia. The subjects are being studied by John Money of Johns Hopkins University and Anke Ehrhardt of Columbia University.

rather than the genetically programmed hormone production that determines sexual orientation and gender role.

CAN WE CHOOSE THE SEX OF OUR CHILDREN?

Methods of Sex Preselection

Aristotle advised Greek women who wanted male children to lie on their right side after intercourse, "for there is the greatest generative heat which is the chief procuring cause of male children." In order to conceive a female child, the woman was advised to lie on her left side postcoitally and to think strongly of a female.

Folk methods have also been used to select a child's sex before or at conception. Coitus has been timed according to wind direction, rainfall, temperature, phases of the moon, and change of tide. Another folk method advised prospective parents to eat sweet foods if they wished to produce a girl and bitter or sour foods for a boy. Because record-keeping has been inadequate or nonexistent we don't know if any of these methods worked.

In different cultures, at different times, women have been under considerable pressure to produce babies of one sex or another (usually male), in spite of the fact that we now know (Fig. 2-4) that it is the male contribution that determines the sex of the child. The theoretical ratio of 1:1 male to female at conception is termed the primary sex ratio. This ratio would be altered if anyone could control, predict, or otherwise guarantee that the fertilizing spermatozoa would be either X or Y.

Timing of coitus and the use of external substances are among the possible ways currently being investigated of guaranteeing "the sex of your choice." The methods for sex preselection now being tested are fairly new, but the idea itself, as we have seen, dates from antiquity. Timing of intercourse with respect to ovulation, changing conditions in the female reproductive tract so as to favor the survival of one type of sperm over another, separating X- and Y-bearing spermatozoa in "test tubes," and determining the sex of the fetus in utero are modern-day possibilities. Do they work?

Evaluation of Sex Preselection Methods

TIMING INTERCOURSE IN RELATIONSHIP TO OVULATION Retrospective data (obtained after the fact) have been assembled to indicate that the conception of males is favored when intercourse takes place some days before the expected time of ovulation (Guerrero, 1974). These same data indicate that artificial insemination done close to the time of ovulation also favors the birth of males.

Timing ovulation can only be done indirectly. Ovulation is accompanied by physical changes in our bodies and these changes can be measured. Some changes are difficult to measure because they require expensive equipment and skilled technicians; others can be measured by anyone with patience and intelligence. One of these latter methods includes the daily measurement of basal body temperature, changes in which are related to ovulation. Some couples have used this method to promote or prevent conception; its theory and practice are explained in Chapter 6.

Whether or not couples wishing to conceive a child of one sex or another will use this method on a prospective basis is debatable. The method works contraceptively only in the hands of highly motivated, reasonably intelligent people, who may use other contraceptives during the fertile period for added security. It would be very difficult, using the basal body temperature graph on a day-to-day basis, to tell, ahead of time, that a woman will definitely ovulate within the next four days. Although the timing of intercourse with respect to ovulation is more scientific than timing it according to the wind, rain, temperature, or moon phase, at present, it doesn't seem that it is much more successful.

CHANGING THE ENVIRONMENT IN THE VAGINA Douching does not kill spermatozoa and thus should not be considered a contraceptive method. It was proposed in the 1930s that alkaline douching media favored the survival of Y-bearing spermatozoa. This "pH theory" stated that an alkaline environment was best suited for the survival of Y-bearing sperm and that X-bearing sperm survived best in an acid medium.

The theory was tested recently in the laboratory. Sperm from the same ejaculate were washed in alkaline or acid media, then examined. If the pH theory were correct, then more living, Y-bearing spermatozoa should have been found in the alkaline medium, while X-bearing sperm should have shown greater viability in the acid environment. The data, obtained by using staining methods that distinguished between (1) living and dead spermatozoa and (2) X and Y bearing spermatozoa, showed that there was no differential survival rate in either medium (Downing, 1976). This method, for the present, should probably be placed with the "sweet for a girl, sour for a boy" folk methods of sex selection.

SEPARATING X- AND Y-BEARING SPERMATOZOA IN THE TEST TUBE "In the test tube" is a general way of saying the sperm is separated in the laboratory from semen specimens collected beforehand. One laboratory method involves the use of a centrifuge. Centrifugation separates substances on the basis of their weight. Heavier particles go to the bottom of the collecting vessel; lighter ones to the top. Persons who believe that this technique can be used for sex preselection maintain that X-bearing spermatozoa are heavier because they have more chromosomal material than Y sperm. They also assume that, if you want to try this method, you (1) will not object to conceiving a child by artificial insemination, and (2) will not hold the laboratory responsible if children of the desired sex, conceived by this method, develop abnormally.

Similar assumptions would be made by other laboratories guaranteeing the same results, if they use other methods of separating X- and Y-bearing spermatozoa. These other methods could involve taking advantage of the theoretical faster movement of Y-bearing sperm (they are lighter and can move faster) or the slight electrochemical difference in charge between the two types of sperm. We may hear more in the future about the different methods of separating the two types of spermatozoa; centrifugation, differential motility, and electrophoresis. A wait-and-see attitude seems justified.

DETERMINING FETAL SEX IN UTERO We have described amniocentesis as a method for diagnosing prenatal genetic defects. Because it can also tell us the sex of the child, it could be used as a method of sex preselection. However, like the amniocentesis done for other reasons, the procedure can only be done in the mid-trimester of pregnancy (cells taken at 15–16 weeks; diagnosis ready at 18–19 weeks). At that time induced abortions are no longer simple or without maternal risk. It is unlikely, therefore, that amniocentesis will be used solely for the purpose of sex preselection. As a matter of fact, only 5 percent of physicians recently questioned on the subject (Rinehart, 1975) indicated that they would be willing to perform an amniocentesis, and possibly, a mid-trimester abortion, solely to guarantee that "It's a girl!" or "It's a boy!"

Sex preselection is considered by some persons to be a form of genetic engineering, which, as such, may be acceptable or unacceptable according to the individual's value system. At this time it is not yet practical on a routine basis; however, the genetic sex of our children may be something we can control in the future. If we could, would we want to?

REFERENCES

OUR GENETIC INHERITANCE

Milunsky, Aubrey. *Know Your Genes.* Boston: Houghton Mifflin Co., 1977. An excellent, authoritative, and well-written book. It should be read by anyone with a personal or professional interest in human genetics.

Littlefield, John W.; Julian N. Kaufer; Edwin H. Kolodny; Vivian Shih; and Leonard Atkins. "Prenatal Genetic Diagnosis," *New Eng. J. Med.* 283 (1970): 1370–81, 1441–47, 1498–1504. Reprinted by the National Foundation,

March of Dimes (Box 2000, White Plains, N.Y. 10602). Limited numbers can be obtained free if they are still available.

Omenn, Gilbert S. "Prenatal Diagnosis of Genetic Disorders," *Science,* May 26, 1972, pp. 952–58.

Stern, Curt. *Principles of Human Genetics.* San Francisco: W. H. Freeman and Company Publishers, 1973.

OUR DEVELOPMENT AS GENETIC FEMALES

Mittwoch, Ursula. *Genetics of Sex Differentiation.* New York: Academic Press, 1973.

Money, John, and Anke Ehrhardt. *Man and Woman: Boy and Girl.* Baltimore: Johns Hopkins University Press, 1972.

Ohno, Susuma. "A Hormone-like Action of H-Y Antigen and Gonadal Development of XY/XX Mosaic Males and Hermaphrodites," *Human Genetics* 35 (1976): 21–25.

———."The Role of H-Y Antigen in Primary Sex Determination," JAMA 239 (January 16, 1978): 217–20.

GENETICS AND GENDER ROLES

Eisinger, A. J.; R. G. Huntsman; Jenny Lord; J. Merry; P. Polani; J. M. Tanner; R. H. Whitehouse; and P. D. Griffiths. "Female Homosexuality," *Nature* 238 (July 14, 1972): 106.

Friedman, R. C.; Inge Dyrenfurth; Daniel Linkie; Ruth Tendler; and Joseph L. Fleiss. "Hormones and Sexual Orientation in Men," *Am. J. Psychiat* 134, no. 5 (May 1977): 571–72.

Gartrell, Nanette K.; D. Lynn Loriaux; and Thomas N. Chase. "Plasma Testosterone in Homosexual and Heterosexual Women," *Am. J. Psychiat.* 134, no. 10 (October 1977): 1117–18.

Griffiths, P. D.; J. Merry; Margaret C. K. Browning; A. J. Eisinger; R. G. Huntsman; E. Jenny A. Lord; P. E. Polani; J. M. Tanner; and R. H. Whitehouse. "Homosexual Women: An Endocrine and Psychological Study," *J. Endocrinol.* 63 (1974): 549–56.

Imperato-McGinley, Julianne; Ralph E. Peterson; Teofilo Gautier; Erasmo Sturla; Elsa Gonzales; and Santiago Pena. "The Impact of Androgens on the Evolution of Male Gender Identity," Abstract from the *Proceedings of the 60th Annual Meeting of the Endocrine Society,* Chicago, June 1977, p. 291.

Kolata, Gina Bari. "Sex Hormones and Brain Development," *Science* 205 (September 7, 1979): 985–87.

Speroff, Leon. "Is There a Biologic Basis for Homosexuality?" *Contemporary Ob. Gyn.* 12 (August 1978): 65–74.

CAN WE CHOOSE THE SEX OF OUR CHILDREN?

Downing, Douglas C. "The Use of a Sequential Staining Technique to Demonstrate the Effect of pH on Survival of Human Spermatozoa." Master's thesis, University of Massachusetts at Amherst, 1976.

Guerrero, Rodrigo. "Association of the Type and Time of Insemination within the Menstrual Cycle with the Human Sex Ratio at Birth," *New Eng. J. Med.* 291 (November 14, 1974): 1056–59.

James, W. H. "Coital Rate, Cycle Day of Insemination and Sex Ratio," *J. Biosoc. Sci.* 9 (1977): 183–89.

Rinehart, Ward. "Sex Preselection: Net Yet Practical," *Population Reports,* series 1, no. 2. Washington, D.C.: George Washington University Medical Center, Department of Medical and Public Affairs, May 1975.

3
Puberty and Sexual Maturation

In endocrine terms, the difference between the child and the adult is quantitative rather than qualitative. (Winter et al., 1978)

As genetic females we grow and develop through childhood up to a point when our female hormones exert their influence on our bodies. This time in our lives is called *puberty,* which has been defined in a biological way as the period in which the organism becomes sexually mature. This period is marked by the occurrence of those constitutional changes whereby the sexes become fully differentiated. At this time the secondary sex characteristics first become conspicuous and the essential organs of reproduction increase in size.

Perhaps we can remember no other period in our lives when "being different" was so traumatic. Peer group comparisons and discussions were often the only source of information we had about the changes that were occurring in our bodies. Some of the changes, for example, the first menstrual bleeding or *menarche,* could have been very frightening if we were unprepared for them. In *adolescence* we are in the midst of all the physiological and psychological changes that occur from puberty to maturity. (Most of these changes occur during our teen years; the terms "adolescents" and "teen-agers" are often used synonymously.)

LOOKING LIKE FEMALES: OUTSIDE AND INSIDE

Adolescence was, for many of us, a worried time. Our bodies were changing either too soon or too late. Our group of female friends with whom we could compare ourselves was probably small enough

TABLE 3–1 Mean Age at Onset of Female Pubertal Changes in the United States

Age	*Characteristic*
9–10	Beginning of height spurt Growth of bony pelvis Female contour fat deposition Budding of nipples
10–11	Budding of breasts Appearance of pubic hair (may precede breast budding in 10% of cases)
11–12	Appearance of vaginal secretions Growth of internal and external genitalia Increase in vaginal glycogen content; lowering of PH
12–13	Pigmentation of areolae Growth of breasts
13–14	Appearance of axillary hair Increase in amount of pubic hair Acne (in 75–90%) Menarche
15–16	Arrest of skeletal growth

Source: Alvin F. Goldfarb, "Puberty and Menarche," *Clin. Obst. Gyn.* 20, no. 3 (September 1977): 629. Reprinted by permission of Harper & Row, Publishers, Inc. and Alvin F. Goldfarb, M.D.

FIGURE 3-1 Factors which influence development of sexual function in the female.

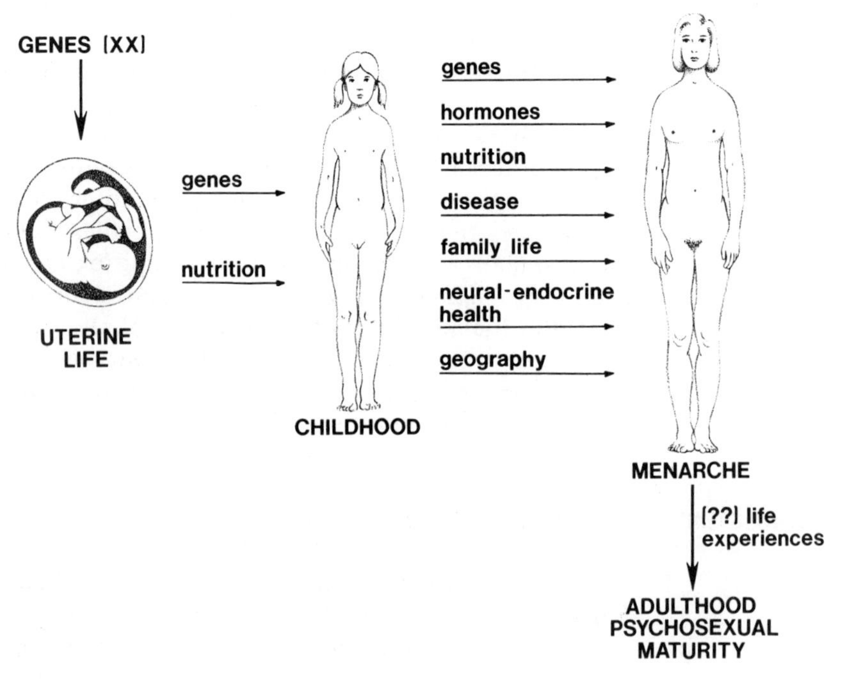

that we had a hard time telling if we were progressing too quickly or too slowly in our developing femaleness. If we had had access to a larger peer group, we could have formed a better judgment of our own pubertal growth. The mean age at which females in the United States experience growth changes of puberty are listed in Table 3-1. The standard deviations from the means are not given in this table, but they would probably indicate an even wider chronological range.

We experienced menarche and the other physical changes of puberty at different times because we are genetically different from our peers. Our genes programmed us to develop on a certain time scale during intrauterine life and throughout childhood. Between childhood and puberty/menarche, other influences—including nutrition, hormones, general health, and geography—were set into the program (Fig. 3-1). These influences are fully expressed when we achieve adulthood. Our life experiences then interact with all the physical factors in the development of adult psychosexual maturity, which, for most of us, is an ongoing life process.

Breast Development

Breast development occurs in stages, which are summarized pictorially in Figure 3-2. These stages of breast development have been described by Tanner (1969) following a longitudinal study of growth in several hundred adolescent girls (Fig. 3-2). The elevation of the nipple in Stage 1 (not shown in Figure 3-2) is followed by the projection of mammary gland tissue from the chest wall (Stage 2). Further increase in the glandular mass occurs in Stage 3, which is also characterized by pigmentation of the areola, the area surrounding the nipple. Stage 4 marks the projection of the areola and nipple to form a secondary mound above the enlarging breasts. The secondary mound formation does not occur in all individuals. In Stage 5 the areolae recede to the general contour of the breast and the nipple only projects. In some women this stage is reached only after

FIGURE 3-2 Breast development. Drawn from Plate 6, J. M. Tanner, *Growth at Adolescence,* 2nd ed. (Oxford: Blackwell Scientific Publications, 1969). By permission of J. M. Tanner.

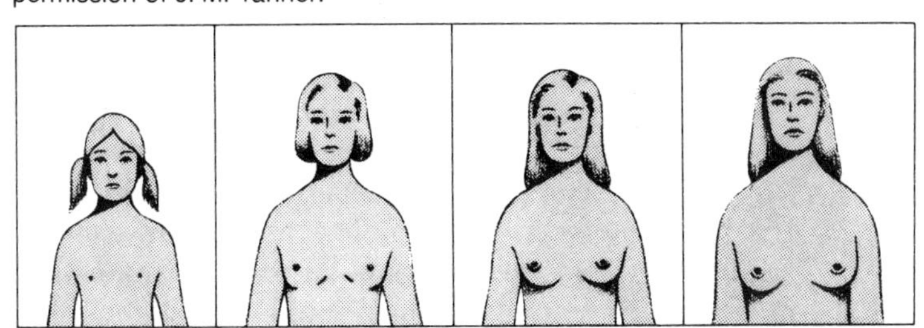

TABLE 3–2 Stages of Female Breast Development

Stage	Description	Age at Onset (Years) Mean	Range
1	Prepubertal; elevation of nipple		
2	Breast bud stage; elevation of breast and nipple as a small mound; enlargement of areolar diameter	11.2	9.0–13.3
3	Further enlargement of breast and areola, with no separation of their contours	12.2	10.0–14.3
4	Projection of areola and nipple to form a secondary mound above the level of the breast (may not occur in all individuals)	13.1	10.8–15.3
5	Mature stage; projection of nipple only, due to recession of areola to the general contour of the breast	15.3	11.9–18.8

Source: Adapted from data in J. M. Tanner, *Growth at Adolescence,* 2nd ed. (Oxford: Blackwell Scientific Publications, 1969). By permission of J. M. Tanner.

a pregnancy. The mean age at which each stage of breast development occurs is indicated in Table 3-2. We should note the wide range around the means. All the factors indicated in Figure 3-1 also influence the time scale for these changes.

A preoccupation with breast size seems to characterize many concepts of ideal female body proportions. It may be well to remind ourselves that the ideal size, contour, and proportion of the female breasts vary from age to age and from time to time within a culture; see, for example, the Venus de Milo and the Playmate of the Month. The size of the breasts has nothing to do with their function, which is baby-feeding.

Although breast development is controlled by estrogen, no direct relationship can be demonstrated between the levels of circulating estrogen and the size of the breasts. It may be that we inherit a sensitivity to estrogen in our target tissues, including the breast, in the form of increased numbers of estrogen receptors. This may be small comfort to the adolescent who thinks that her breast size is forever stuck at Stage 3; however, the growing psychosexual maturity of emerging adulthood may make her more comfortable about her body contours, including breast size.

Pubic Hair

The growth of pubic hair is also a secondary sex characteristic; it grows in a triangular pattern in the male and flat-topped across the mons pubis in the female. Our initial growth of pubic hair and axillary hair is the result of androgen (male hormone) stimulation on the part of the adrenal glands. Estrogen influences the female

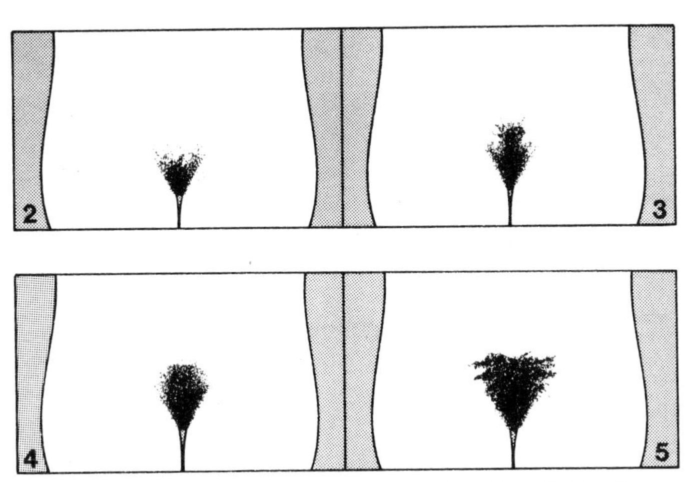

FIGURE 3-3 Pubic hair development. Drawn from Plate 7, J. M. Tanner, *Growth at Adolescence,* 2nd ed. (Oxford: Blackwell Scientific Publications, 1969). By permission of J. M. Tanner.

growth pattern, which is depicted in Figure 3-3. In Stage 1 (not shown) there is no hair. With the onset of puberty the pubic hair develops as follows:

Stage 2: Sparse hair, only lightly curled, on the labia majora.

Stage 3: Hair darker, curlier, spread over mons pubis.

Stage 4: Dark coarse hair covers mons pubis in adult pattern.

Stage 5: Hair extends over inner surface of thighs.

The mean age at which each of these stages occurs is shown in Table 3-3.

TABLE 3–3 Stages of Female Pubic Hair Development

Stage	*Description*	*Age at Onset (Years)* *Mean*	*Range*
1	Prepubertal—no hair		
2	Sparse, long, slightly pigmented downy hair, straight or only slightly curled; appears chiefly among the labia	11.7	9.3–14.1
3	Considerably darker, coarser and more curled; spreads sparsely over mons pubis	12.4	10.2–14.6
4	Curled, coarse hair covering mons pubis but not onto thighs	13.0	10.8–15.1
5	Hair of adult quantity and type with spread to medial surface of thighs	14.4	12.2–16.7

Source: Adapted from data in J. M. Tanner, *Growth at Adolescence,* 2nd ed. (Oxford: Blackwell Scientific Publications, 1969). By permission of J. M. Tanner.

The Adolescent Growth Spurt

Many of us can remember (with family photographs to bolster our memories) that we were almost as tall as we are now at about age 12 or 13, and that most of the boys in our seventh and eighth grade classes were too short to be interesting, much less to be dance partners. If we look at Figure 3-4, we can see that gains in height per year occur at a peak rate about two years earlier in girls than in boys.

The growth spurt that we experienced as adolescents involved changes in our skeletal system. Rising hormone levels stimulated bone cells that replaced cartilage at the ends of the long bones. When all the cartilage had been replaced by bone, growth of the long bones, hence, growth in height, was over. This arresting of skeletal development becomes a consideration when estrogens in the form of contraceptive steroids are ingested by very young teen-agers.

If we refer again to Figure 3-4, we can reaffirm that we gained weight faster and earlier than our male counterparts. Fat deposition, one of the effects of estrogen, is part of puberty. Fat on the hips, buttocks, and breasts enhances our female secondary sex characteristics—a mixed blessing if too much fat gets deposited. In boys, on the other hand, the male hormone, testosterone, stimulates the

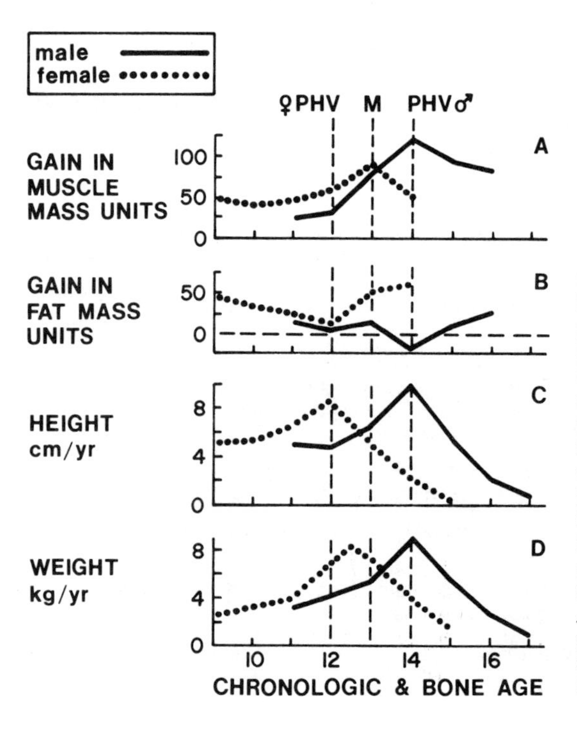

FIGURE 3-4 Changes in height and weight during puberty. Mean changes in muscle mass, fat mass, height, and weight as correlated to stage of pubertal development and mean chronologic age in normal males and females. PHV is peak height velocity. From H. V. Barnes, "Physical Growth and Development during Puberty," *Med. Clin. North Am.* 59 (1975): 1305. By permission of W. B. Saunders Company Publishers.

development of muscle rather than fat. At an average age of 14, the deposition of fat and muscle are occurring at opposite rates in girls and boys.

Development of External Genital Structures

Before puberty our hairless external genitalia, which are called collectively the vulva or pudenda, face forward. Pubertal development involves both the development of hair and the downward facing of the vulva. (The component parts of the vulva include the labia majora and minora, the clitoris, the mons pubis, the orifice and vestible of the vagina, and the glands associated with the vagina.) At puberty the labia majora and minora enlarge to close the entrance to the vagina. The erectile system of the clitoris develops and fat is deposited on the mons pubis. The glands in the vestibule of the vagina, which produce a secretion during intercourse, also develop in size during puberty.

The pre- and postpubertal development of our external genital structures are depicted in Figure 3-5. The hymen, which is illustrated in position in the lower portion of the figure, consists of a thin layer of mucous membrane near the opening of the vagina. The hymen is broken during the first coital activity, if not before. Although it has no known function, the observation and/or proof of an intact hymen has been a condition for a woman's acceptability as a marriage partner in some cultures.

Development and Growth of Internal Reproductive Structures

Our ovaries increase in volume during puberty, while uterine size doubles and even triples. In our oviducts, or fallopian tubes, two types of structures grow in size and increase in number. Smooth muscle and cilia develop during puberty in preparation for reproductive maturity. After ovulation, the ovum moves down to the site of fertilization, propelled by the cilia and smooth muscle.

Thus the development of internal and external structures that are related to sexual and reproductive functioning are the primary events of puberty. We have seen that the occurrence of some of these events has been divided into stages. The stages or changes are readily demonstrable as our bodies mature. The hormonal mechanisms that control the onset and progress of puberty have been somewhat clarified recently, due to the development of sensitive measuring systems described in Chapter 1. However, the precise hormonal changes and cause-and-effect relationships of the menarche remain to be defined.

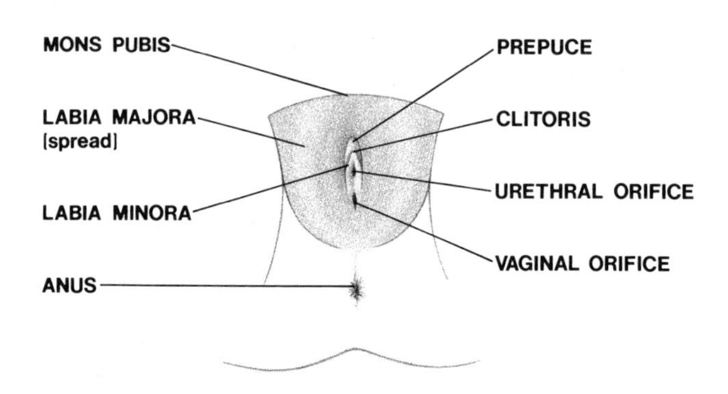

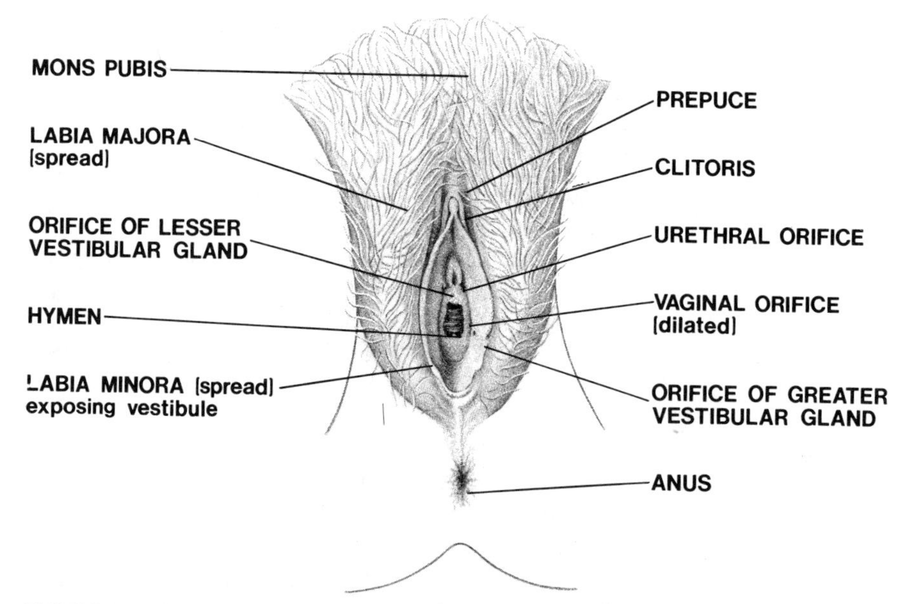

FIGURE 3-5 Development of the external genital structures (vulva). The upper drawing represents the genital structures in the prepubertal girl. The growth of pubic hair and the development of the vulva occur during puberty.

PUBERTY AND MENARCHE: HORMONAL CHANGES AND CAUSE-AND-EFFECT RELATIONSHIPS

Puberty Redefined

Another way of looking at puberty is to examine the changing hormone levels. In these endocrine terms, puberty can be redefined as the occurrence of physical signs (those described above) attributable to increased blood levels of sex steroids. These steroids, in turn, are associated with rising blood levels of follicle-stimulating hormone (FSH) and luteinizing hormone (LH). Before puberty these gonadotropin levels are lower and can be suppressed by smaller amounts of steroids than after puberty.

In other words, all the external changes that befall us during puberty are directed by increases in our hormone levels. Figure 3-6 illustrates the secretion of ovarian hormones (estrogen and progesterone) and gonadotropic hormones (FSH and LH) in preadolescent, mid-pubertal and postmenarchal girls. Preadolescent girls secrete very low levels of both the ovarian and the gonadotropic hormones (A). By mid-puberty (B) the ovaries are secreting estrogen in a rhythmic fashion, and these higher levels of estrogen begin to exert some feedback effect on FSH secretion. The maturing ovarian follicles, which produce the estrogen, respond to stimulation by FSH. After ovulation the ovary produces progesterone. This initiates the menarchal and postmenarchal pattern of hormone secretion (C).

Fetal Endocrine Sexual Development

We have already seen that sexual development in males requires the Y chromosome and the H-Y antigen. In contrast, the differentiation of female sex organs proceeds without hormone stimulation during intrauterine life. Our fetal pituitary glands produce some LH and FSH during the second trimester of fetal life and these hormones stimulate the development of our ovaries.

During the seventh month of fetal life, something happens in

FIGURE 3-6 Hormone patterns during puberty in girls. Note increasing levels of FSH and LH as puberty progresses. The onset of rhythmic FSH fluctuations and establishment of adult patterns (C) begin some time after menarche. Progesterone values do not change markedly until ovulation begins. From C. Faiman, J. S. D. Winter, and F. I. Reyes, "Patterns of Gonadotrophins and Gonadal Steroids throughout Life," *Clin. Obstet. Gynecol.* 3 (1976): 467. With permission of the publisher, W. B. Saunders Company Ltd.

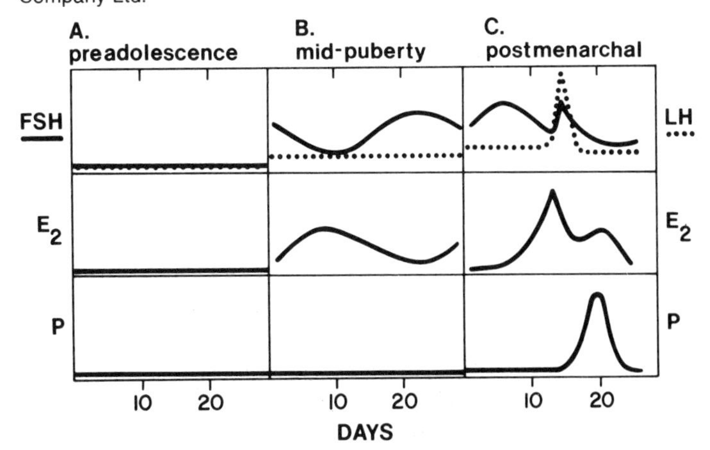

our ovaries for which there is no explanation. Some of the hundreds and thousands of prospective ovarian follicles developing in the tiny ovaries do not mature; they regress. These primordial germ cells number about 10 million during the fifth month of fetal life. The regression that commences during the seventh month reduces the total number of prospective ova to about 2 million at birth. This phenomenon of follicular atresia, or regression of cells destined to become ova, continues after birth, and so we, as females, have a plentiful but limited supply of prospective ova as we pass through the various stages of our reproductive lives.

Hormone Levels during Childhood

The secretion of pituitary and ovarian hormones is low during childhood. There is, however, some secretion of steroid hormone in the form of androgens from the adrenal glands. Before puberty, these androgenic hormones promote the growth of axillary and pubic hair and also influence the adolescent growth spurt. They are also implicated in something we would rather forget about: acne. Some recent dermatological studies (Pocchi et al., 1977) have shown that androgen-directed changes in the surface lipid composition of the skin may be one of the earliest signs of puberty. These lipids are secreted by androgen-stimulated sebaceous glands. At the same time, the excretion of urinary androgens metabolites may be positively correlated with age and sometimes with acne in prepubertal girls. At least we can blame something (androgens) for the distresses of our adolescent acne!

Although models that would fully explain how androgens, which are male hormones, affect our female development are not available, the diagram in Figure 3-7 indicates a likely endocrinological sequence. It is possible that external influences in our environment, operating through the hypothalamus, initiate the release (1) of corticotropic-releasing factor (CRF). The pituitary is then stimulated to produce (2) adrenocorticotropic hormone (ACTH), which affects the adrenal cortex. The weak androgens produced (3) by the adrenal cortex stimulate the growth of axillary and pubic hair, the growth spurt, and lipid secretion by the epidermal sebaceous glands.

Hormone Levels during Puberty

Breast development, female fat distribution, skeletal maturation, and vaginal and uterine development are, as we have seen, the results of the increasing amounts of estrogen produced by the ovary during late puberty. At some time during late puberty, the maturing ovarian follicles respond to FSH from the pituitary. The estrogen and FSH begin to operate as a negative feedback system; high levels of one

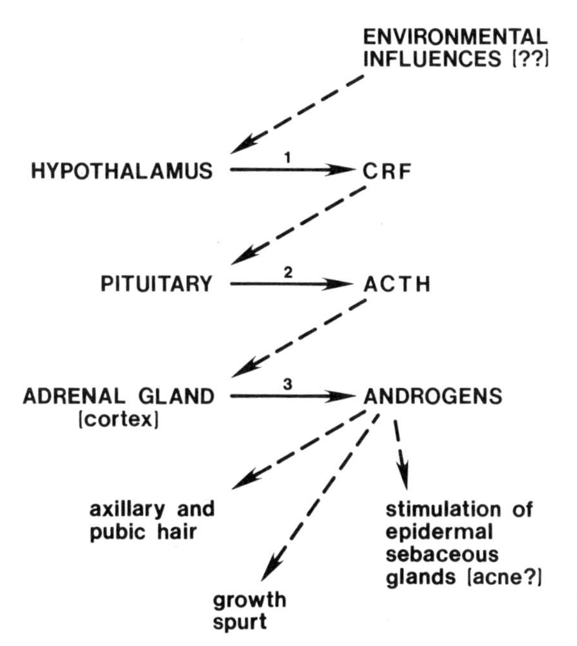

FIGURE 3-7 Androgens and puberty.

"shuts off" production of the other, and a cyclical hormone production pattern emerges. We experience our menarche or first menstrual bleeding when estrogen levels are lowered toward the end of the newly established cycle. The lining of the uterus is shed because it is no longer sustained by the hormonal stimulation of estrogen.

Our early adolescent menstrual cycles are usually not ovulatory cycles; that is, an ovum is not released from the follicle into the oviduct. As our ovaries produce more and more estrogen, LH is released from the pituitary with the FSH and these two hormones together trigger the rupture of the ovarian follicle (ovulation). After ovulation, the corpus luteum in the ovary produces progesterone, and postmenarchal ovulatory cycles are established, as shown in Figure 3-6(C). Our adult menstrual (ovulatory) pattern of hormone secretion thus appears after an evolutionary sequence of developmental stages that can vary greatly among individuals.

THE ONSET OF PUBERTY AND THE EXPERIENCE OF MENARCHE

In the absence of definite and precise hormonal mechanisms to associate with the onset of puberty and menarche, several different explanations have been proposed. These include:

1. *The gonadostat theory (Grumbach et al., 1974)*
2. *Biorhythm theory (Weitzman et al., 1975)*
3. *Critical body weight theory (Frisch, 1974).*

If these explanations are combined, they indicate that puberty and menarche occur, not because of one factor, but because of several interrelated and complex factors.

THE GONADOSTAT THEORY The gonadostat theory proposes that the release of gonadotropins from the pituitary gland is inhibited by the low levels of steroid secretion in our prepubertal gonads. As the central nervous system matures at the end of childhood, the hypothalamus responds by secreting gonadotrophin-releasing factor (GRF). During puberty and adulthood, GRF stimulates increases of LH and FSH, followed by increased estrogen production. The adult pattern of hormone secretion involves a decrease in the insensitivity of the gonadostat (or an increase in its sensitivity). See Figure 3-8.

BIORHYTHM THEORY Biorhythms are associated with changing activities in our bodies according to the time of day, of month, and of year. One theory for the onset of puberty relates the increase in sleep time of preadolescents to increases of LH and FSH in their plasma. This may begin in childhood. It is possible that increases in sleep periods stimulate increased secretion of gonadotropins via the central nervous system and the hypothalamus.

CRITICAL BODY WEIGHT THEORY A nervous system synchronization of hypothalamic reproductive control based on the concept of a critical body weight has been proposed. This theory suggests that the fat/lean or fat/body weight ratios act as triggers for the onset of reproductive function in the female. Evidence to support this theory comes from several sources:

1. *Urinary and plasma gonadotropins are low in underweight patients.*
2. *The response to GRF is lessened when body weight is low, and it is restored when body weight returns to normal.*
3. *The secretory "biorhythm" pattern of LH secretion before and during puberty is closely correlated with the extent of gain or loss of body weight (Frisch and McArthur, 1974).*

The critical body weight hypothesis has been the subject of some debate (Wilen and Naftolin, 1977); nevertheless the available evidence suggests some association between body mass/composition and pubertal development.

It is possible that we experienced our first menstrual periods when our genetically determined pattern of hormone secretion was somehow correlated with our size and body weight. The proper amount of rest possibly conditioned the central nervous system to

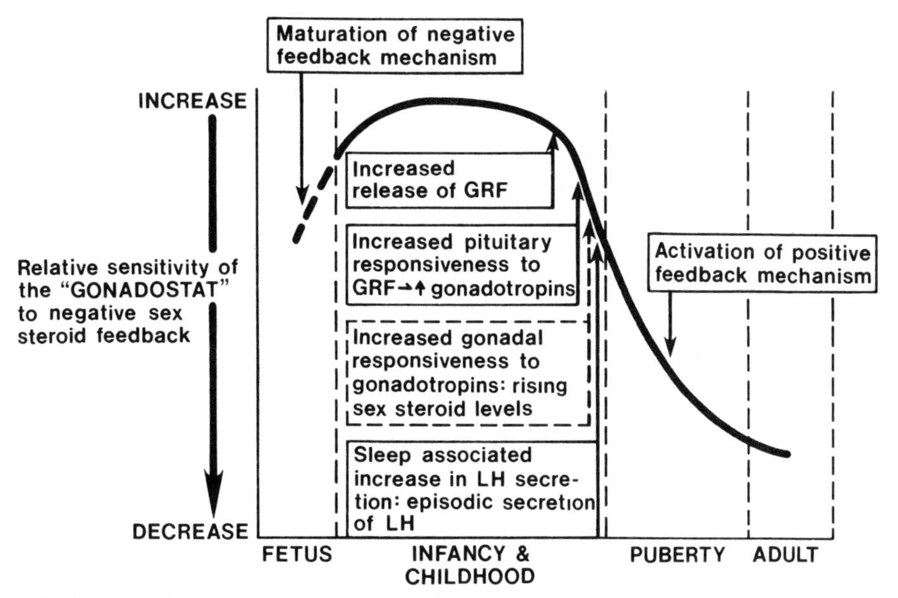

FIGURE 3-8 The gonadostat. From M. M. Grumbach, J. E. Roth, S. L. Kaplan, and R. P. Kelch, "Hypathalamic-Pituitary Regulation of Puberty: Evidence and Concepts Derived from Clinical Research," in *Control of the Onset of Puberty,* ed. M. M. Grumbach, G. D. Grave, and F. E. Mayer (New York: John Wiley & Sons, Inc., 1974) pp. 115–66. By permission of John Wiley & Sons, Inc.

respond to hormonal as well as nervous stimuli in the establishment of the cycles. Problems as to *when* puberty and menarche occurred in our lives may have been forgotten; however, some of the menstrual problems we encountered as adolescents may be with us still.

MENSTRUAL PROBLEMS OF THE ADOLESCENT

Many of us have experienced menstrual dysfunctions during the first year or so after menarche. For the most part, we outgrew them, but for some of us these were manifestations of chronic reproductive disorders that still bother us to some extent.

Precocious Puberty Precocious puberty is defined at the present time as the occurrence of menses before the age of 10. While precocious puberty may be upsetting and may require some special counselling for the individual girl, this condition indicates no abnormality in 80 to 90 percent of girls in whom it occurs. It is a genetically determined characteristic, like the other manifestations of puberty. In some of the remaining cases, ovarian tumors that produce excess estrogen may stimulate the gonadostat as described above. Other cases of precocious puberty involve lesions of the central nervous system. Individuals who experience precocious puberty can (and often do) become mothers at *very* early ages.

AMENORRHEA Amenorrhea is the failure of menses to occur within two and a half years after breast and pubic hair development. This failure to menstruate may be a result of genetic disorders of sex determination, as described in Chapter 2. Faulty development of the genital tract, uterus, or vagina, or imperforate hymen may also be implicated in amenorrhea. The tendency of teen-agers to embrace crash or fad diets should also be considered in view of the evidence for critical body weight and the onset of menses.

ANOREXIA NERVOSA Anorexia nervosa is an extreme example of nutrition-related amenorrhea. This syndrome occurs chiefly in adolescent girls who may have neuroses related to food in the absence of any structural disease. The aversion to food may be motivated by an abnormal desire to become thin or by fear of gastrointestinal consequences of food ingestion. Extreme emaciation in these girls may need to be treated by tube feeding, accompanied by skilled counselling so that the mental attitude of the patient can be altered. If the condition is cured through diet and counselling, menstrual cycles recommence following weight gain. Unfortunately, some cases do not respond to therapy and patients die from starvation and/or infection.

MENSTRUAL CRAMPS Menstruation-associated cramps are an example of an adolescent reproductive disorder that may persist into adulthood. The experience of cramping usually does not begin until menstrual cycles are ovulatory, that is, until progesterone is being produced. This occurs during the postmenarchal period shown in Figure 3-6(C). It is presently believed that substances called prostaglandins, produced in the endometrium, are the primary causes of menstrual cramps. Prostaglandin antagonists, such as aspirin and amphetamine, alleviate the symptoms of menstrual cramps. Progestogens are also effective when given five to seven days before the expected onset of menses.

It is not necessary for us (even if we had the luxury of being able to do so) to treat ourselves as semi-invalids during the bleeding phase of our menstrual cycles. Tests on groups of women who were experiencing dysmenorrhea showed that their physical performance was not adversely affected during menstruation. In the absence of any other organic disturbance, parameters of physical performance such as heart rate and oxygen uptake did not change during different phases of the menstrual cycle. Pulmonary ventilation was increased and reaction time slightly decreased, but this did not seem to affect the capacity of the women for work (Gamberale et al., 1975).

DYSMENORRHEA Some of us probably have our own categories of special aches and pains that we could add to a list of woes that are frequently a part of the menstruation experience: cramps, headache, backache, legache, breast fullness or tenderness, abdominal swelling, and nausea. Our experience of these symptoms of dysmenorrhea may be partly our own fault. Some gynecologists estimate that from 70 to 80 percent of our experiences of painful menstrual symptoms are due to faulty living habits which include lack of exercise, poor posture, fatigue, irritability, and tension. Our own attitudes may also intensify menstrual problems; we may not understand what menstruation really is and we may not have a healthy, wholesome attitude toward it. A recent study of the beliefs of young adolescents concerning menstruation compared their premenarchal attitudes with those of older girls who had established menstrual cycles (Clarke and Ruble, 1978). The young adolescents recorded surprisingly negative attitudes and expectations, and the nature and timing of information received from others was implicated in these negative feelings. If some of us are parents, educators, or health science professionals, we need to be good communicators so that young girls are adequately informed of the social and psychological, as well as the physiological consequences of reproductive maturity.

Dysfunctional Uterine Bleeding Dysfunctional uterine bleeding may occur during the menopause, when the negative feedback systems that control periodicity are overridden by diminished estrogen production. In the adolescent, the same symptom, unusually heavy menstrual bleeding, generally proceeds from the opposite cause. Dysfunctional uterine bleeding occurs because of the unopposed effect of excessive amounts of estrogen. Large amounts of estrogen inhibit FSH and LH release, and persistent anovulatory cycles succeed one another. Absence of ovulation is the usual cause of dysfunctional uterine bleeding, which is defined as any menses lasting more than eight days with a blood loss of more than 150cc. Certain other reproductive complications can mimic the failure of ovulation resulting in dysfunctional uterine bleeding. These include complications of early pregnancy, adenocarcinoma of the vagina or cervix, the presence of an intrauterine device (IUD), misuse of contraceptive pills or other medication, narcotic abuse, ovarian neoplasia, cervical or endometrial polyps, and general disorders of the endocrine system.

HEALTH NEEDS OF THE ADOLESCENT

The physiological events that occur during puberty are certainly not without a stress component. An environment that is helpful and supportive can assist the adolescent girl to grow psychologically by coping successfully with stresses that result from physiological and social pressures. In other words, the more the environment of childhood and adolescence offers reasonable support and guidance, the better the adolescent can cope with immediate and future stress. New and accurate information can convert an uncertain situation into a predictable one. This security enables the adolescent to cope with stresses in the following areas:

1. *the experience of menarche*
2. *need for proper rest, nutrition and exercise during adolescence*
3. *experimentation with tobacco, drugs and alcohol*
4. *sexual activity, pregnancy, and venereal disease.*

Preparation for and Support during Menarche

As noted above, the beliefs of young adolescents concerning menarche influence their experience of its physiological aspects (Clarke and Ruble, 1978). Additional evidence from a large gynecological study in Europe suggests that attitudes toward menarche are related to gynecological distress patterns during adult life (Wenderlein, 1977). Does menarche, in fact, represent such an important event for girls during puberty that it affects the experience of later gynecological events in women? Apparently, the answer is yes. An early experience of painful menarche, in which the subjects were surprised by the event because of inadequate preparation, predisposed them toward:

1. *the experience of menstruation as painful*
2. *impaired well-being during pregnancy*
3. *negative consequences after gynecological operations, e.g., mastectomy, hysterectomy.*

It is possible that accurate information about the events that occurred in their bodies during menarche, presented before the event, could have made the transition, and its consequences, less painful.

Establishment of Good Health Habits

In many societies the consequences of malnutrition interfere with the physiological processes of puberty. An extreme example of malnutrition in an affluent society, anorexia nervosa, has been described. Less extreme dietary imbalance may be unimportant for short or medium periods. For example, crash diets do not usually last long enough to deplete the body's energy reserves (see Chapter 11). On the other hand, high-energy, high-fat diets constitute a risk factor for coronary disorders in adult life.

For better or worse, the habits of eating, sleeping, and exercise which may persist into adult life are established during the period of adolescence. Accurate information and instruction may or may not be accepted, but they should be available.

Use of Tobacco, Drugs, and Alcohol

Peer pressure on the adolescent may be so intense that accurate information is disregarded. Insofar as smoking is concerned, studies by the World Health Organization have shown that:

1. *The use of tobacco is most common among young people whose families, friends, or peer groups also smoke cigarettes.*
2. *It is relatively more common among young people whose life objectives are relatively low and who do not do well in school (underachievers).*
3. *For some adolescents, of either sex, smoking is a symbol of independence and/or rebellion against family or group standards.*
4. *The notable health hazards of smoking (cancer, chronic bronchitis, coronary disease, emphysema) are regarded by most young people as presenting merely potential, distant threats. Information about the risks of early respiratory function impairment may make a stronger impression.*
5. *School courses on the subject are often deficient for reasons involving instructors as well as students*

Perhaps the same reasons given above for the use of tobacco by adolescents could be cited for their use of drugs and alcohol. The physiological consequences of alcohol abuse are considered in Chapter 10. The adolescent should be aware of the well-documented effects of excessive alcohol consumption: cirrhosis of the liver, chronic malnutrition, and brain damage. Damage to the fetus of an alcoholic mother (Fetal Alcohol Syndrome) is a possibility for teen-age as well as for adult mothers.

Sexual Activity and Pregnancy

Many socioeconomic factors, as well as the physiological attainment of sexual maturity, contribute to and encourage sexual activity among teen-agers. These include, among others:

1. *the sexuality of contemporary society and its expression in the media*

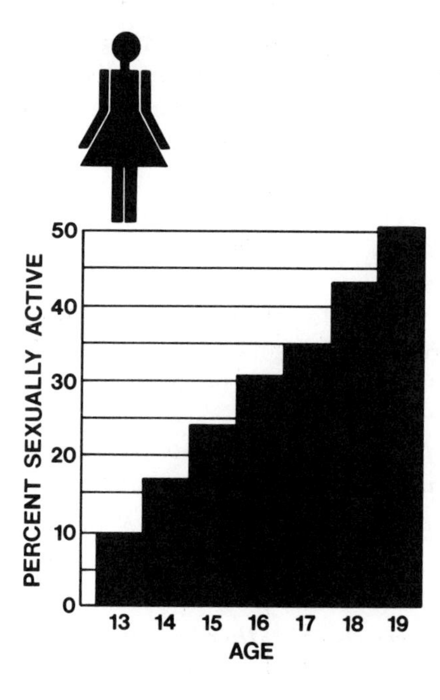

FIGURE 3-9 Sexual activity in U.S. females, 1974–1975. From R. Lincoln, Fred S. Jaffe, and A. Ambrose, *11 Million Teen-Agers* (New York: Alan Guttmacher Institute, 1974), p. 7. By permission of Alan Guttmacher Institute.

2. *peer and social pressure*
3. *psychological and emotional problems of the adolescent.*

Perhaps the most serious contribution by us as adults to the adolescent's use and misuse of her newly acquired sexual capabilities is the conspiracy of silence with which we surround human sexuality. Somehow, our society seems to be unable to admit and to deal realistically with the sexual activity of adolescents. Figure 3-9 represents sexual activity with respect to age in females who were 13 to 19 years old in 1974–75.

The unrealistic attitude of adults is underlined by failures to provide sex education, birth control information, and a set of value judgments against which to judge the appropriateness of sexual activity. Teen-age girls are physiologically prepared for sexual activity long before they are psychologically and economically ready to cope with the consequences of unprotected intercourse, namely, pregnancy.

Few teen-agers begin to use contraceptives when they begin sexual activity. In many cases, their use of contraceptives is sporadic and/or inadequate. The simple fact that early menstrual cycles are usually anovulatory may protect some girls for a while, but not for long. The false idealism that the use of contraceptives detracts from the fullness and spontaneity of the relationship usually has one hard knock with reality: pregnancy.

In 1974 it was estimated that about 1 million girls between the ages of 15 and 19 become pregnant each year. Many girls are much younger than this when they become mothers. About 13 percent of the 1 million pregnancies ended in spontaneous abortions; another 27 percent were terminated by induced abortion. Sixty percent (approximately 600,000) culminated in delivery, one-third of which were out of wedlock. Many "wedlocks" doubtless took place only after the fact of pregnancy was established. Teen-age marriages are two to three times more likely to break up than unions of those who marry in their twenties. Teen-age couples who marry as a result of pregnancy are also more likely to be economically disadvantaged. Many young mothers, married or unmarried, do not complete high school and have no marketable job skills. Their employment is further restricted by child care obligations. The responsibility of caring for another human being and of avoiding future pregnancies seems to be more difficult for these young adolescents.

Unless the girl is very young, the anatomical/physiological complications of pregnancy are not as severe as one would expect. Pregnancy problems of the adolescent—such as anemia, toxemia, and bleeding—may be related more to poverty and inadequate prenatal care than to maternal age. One study (Duenhoelter et al., 1975), which matched patients under 15 years of age (study group) to those between 19 and 25 (control group), found significant differences between pregnancy complications in only a few instances. The study group differed from the control group chiefly in three physiological and one psychological aspect:

1. *They had experienced earlier menarche.*
2. *They had significantly greater occurrences of pregnancy-induced hypertension.*
3. *They presented more cases of pelvic inlet contraction.*
4. *They had a greater number of recurring pregnancies within 18 months of the initial pregnancy.*

Teen-age mothers *given proper care* may have fewer complications in childbirth than was previously expected. However, teen-agers are not as good mothers as older women. The health of their babies, better at birth, deteriorates during the first year, except in cases in which the babies live with their grandmothers.

The teen-age or adolescent period—the time of puberty and menarche in girls—spans the periods involving the following changes:

1. *the changes between the initial appearance of secondary sex characteristics and the achievement of sexual maturity*
2. *the changes in psychological processes and identifications that occur during the transition from child to adult*

3. *the change from total socioeconomic dependence to relative independence.*

The first change will precede the second and the third changes by a considerable time period in most cases. As she reaches maturity, the adolescent should, with the support of those responsible for her well-being, realistically consider *in advance* the implications of possible sexual involvement, including a plan for effective contraception.

REFERENCES

LOOKING LIKE FEMALES

Goldfarb, Alvin F. "Puberty and Menarche," *Clin. Obs. Gyn.* 20, no. 3 (September 1977): 625–31.

Marshall, W. A., and J. M. Tanner. "Variations in the Pattern of Pubertal Changes in Girls," *Arch. Dis. Child.* 44 (1969): 291.

Tanner, James M. "Growth and Endocrinology of the Adolescent," in *Endocrine and Genetic Diseases of Childhood,* ed. Lytt I. Gardner, pp. 19–60. Philadelphia: W. B. Saunders Company Publishers, 1969.

van der Werff ten Bosch, J. J.; Anneke Bot; and B. T. Donovan. "Puberty," *Research in Reproduction* 8, no. 6 (November 1976).

PUBERTY AND MENARCHE: HORMONAL CHANGES AND CAUSE-AND-EFFECT RELATIONSHIPS

Nelson, Ronald M. "Physiologic Correlates of Puberty," *Clin. Obs. Gyn.* 21, no. 4 (December 1978): 1139–49.

Pocchi, Peter E.; John S. Strauss; and Donald T. Downing. "Skin Surface Lipid Composition, Acne, Pubertal Development and Urinary Excretion of Testosterone and 17-Ketosteroids in Children," *J. Invest. Derm.* 69, no. 5 (1977): 485–89.

Winter, Jeremy S. D.; Charles Faiman; and Francisco I. Reyes. "Normal and Abnormal Pubertal Development," *Clin. Obs. Gyn.* 1 (March 1978): 67–86.

THE ONSET OF PUBERTY AND THE EXPERIENCE OF MENARCHE

Frisch, Rose E., and Roger Revelle. "Height and Weight at Menarche and a Hypothesis of Menarche," *Arch. Dis. Child.* 46 (1971): 695–701.

Frisch, Rose E., and Janet W. McArthur. "Menstrual Cycles: Fatness as a Determinant of Minimum Weight for Height Necessary for Their Maintenance and Onset," *Science,* September 13, 1974, pp. 949–51.

Frisch, Rose E. "Critical Weight at Menarche, Initiation of the Adolescent Growth Spurt and Control of Puberty," in *Control of the Onset of Puberty,* ed. M. M. Grumbach, G. D. Grave, and F. E. Mayer, pp. 403–73. New York: John Wiley, 1974.

Grumbach, M. M.; J. E. Roth; S. L. Kaplan; and R. P. Kelch. "Hypothalamic-Pituitary Regulation of Puberty: Evidence and Concepts Derived from Clinical Research," in *Control of the Onset of Puberty,* ed. M. M. Grumbach, G. D. Grave, and F. E. Mayer, pp. 115–166. New York: John Wiley, 1974.

Weitzman, E. D.; R. M. Boyer; S. Kapen; and L. Hellman. "The Relationship of Sleep and Sleep Stages to Neuroendocrine Secretion and Biological Rhythms in Man," *Recent Progress in Hormone Research* 31 (1975): 399–441.

Wilen, Richard, and Frederick Naftolin. "Critical Body Weight and Sexual Maturation," *Res. Reprod.* 9, no. 5 (September 1977): 1.

MENSTRUAL PROBLEMS OF THE ADOLESCENT

Altchek, Albert. "Dysfunctional Uterine Bleeding in Adolescence," *Clin. Obs. Gyn.* 20, no. 3 (September 1977): 633–50.

Clarke, Anne E., and Diane N. Ruble. "Young Adolescents' Beliefs Concerning Menstruation," *Child Development* 49 (1978): 231–34.

Dickey, Richard P. "Menstrual Problems of the Adolescent," *Postgraduate Medicine* 60, no. 4 (October 1976): 183–87.

Gambarale, Francesco; Lotten Strindberg; and Inger Wahlberg. "Female Work Capacity during the Menstrual Cycle: Physiological and Psychological Reactions," *Scandanavian Journal of Work, Environment and Health* 1, no. 2 (June 1975): 120–27.

HEALTH NEEDS OF THE ADOLESCENT

Duenhoelter, Johann H.; Juan M. Jiminez; and Gabrielle Baumann. "Pregnancy Performance in Patients under Fifteen Years of Age," *Obstet. Gynecol.* 46, no. 6 (July 1975): 49–52.

Hunt, William Burr III. "Adolescent Fertility: Risks and Consequences, *Population Reports,* series J, no. 10, July, 1976.

Wenderlein, J. M., "Menarche: Psychosomatic and Psychosocial Aspects for Gynaecology," *Arch. Gynakol.* 223, no. 2 (1977): 99–114.

World Health Organization. "Health Needs of Adolescents." Technical Report Series no. 609. Geneva: World Health Organization, 1977.

———. "Pregnancy and Abortion in Adolescence." Technical Report Series no. 583. Geneva: World Health Organization, 1975.

4
Reproductive Control Systems

For most of us, the onset of menstruation and ovulation marks the beginning of a long period of reproductive capacity which extends to the end of the menopause. In this chapter, we shall focus on the mechanisms that control our own monthly cycles of fertility, as well as those that control fertility in our male partners. We shall conclude by looking at the changes that characterize the menopause, the end of our reproductive period.

THE FEMALE REPRODUCTIVE CYCLE AND ITS CONTROL SYSTEMS

"To every thing there is a season. . . ." The rhythm of life surrounds us—day and night, high tide and ebb tide, growth and decline, birth and death. Rhythm is so much a part of life that we take it for granted. Our cyclic reproductive functions are under a rhythmic control that begins at menarche and ends at menopause (Fig. 4-1). The components of the control system are the hypothalamus, the pituitary gland, and the ovary. The hormones secreted by these glands interact with one another in a negative feedback system. These hormones also act on organs and tissues that are part of the reproductive system. Alternating monthly periods of fertility and sterility characterize the phase of our lives during which reproduction—childbearing—is possible. The periods from birth to menarche and from the climacteric to senescence and death involve gradual gain then loss of reproductive ability.

During late puberty the pituitary gland produces follicle-stimulating hormone (FSH). The immediate effect of FSH on the follicle is that more estrogen is produced. When the estrogen produced by

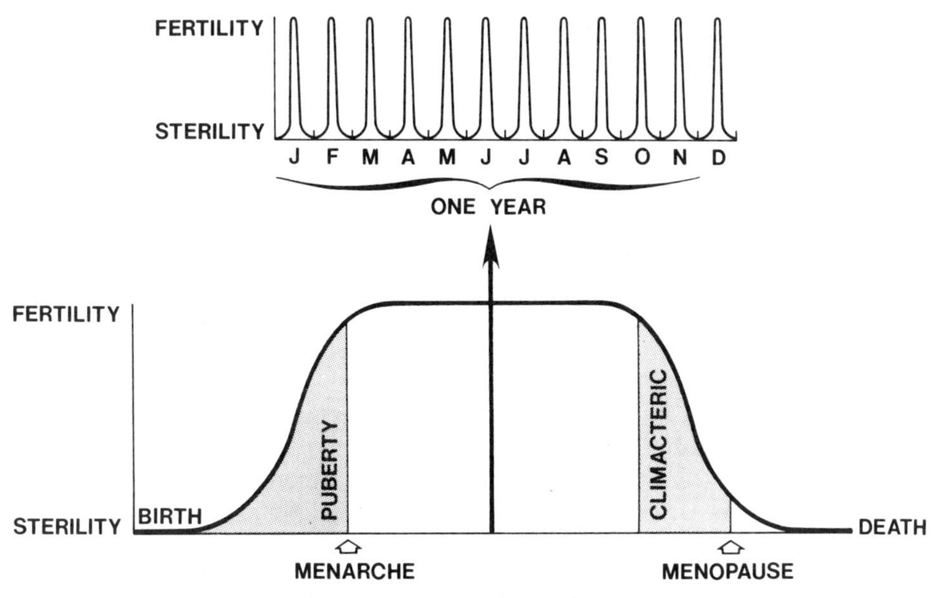

FIGURE 4-1 Reproductive rhythms in our bodies.

the follicle reaches a critical level it has two feedback effects on the pituitary gland. One effect is the inhibition of further FSH production. This completes one component of the cycle: pituitary–FSH–ovary–estrogen–pituitary. Another effect of the estrogen produced by the ovarian follicle is to stimulate the pituitary to begin the production of luteinizing hormone (LH). Luteinizing hormone and FSH work together to stimulate the release of an ovum from the ovary. Luteinizing hormone also acts on the follicle cells to produce progesterone. Progesterone, like estrogen, has a negative feedback effect on the pituitary. When progesterone reaches a critical level it acts on the pituitary to inhibit further LH production. This completes another component of the cycle: pituitary–LH–ovary–progesterone–pituitary.

The role of the pituitary gland in the reproductive cycle is influenced by the hypothalamus. As was illustrated in Figure 1-9, the hypothalamus is located just underneath the middle of the brain, below the thalamus and above the pituitary gland. At the beginning of a new reproductive cycle, gonadotropic releasing factor (GRF) is produced by neurosecretory cells in the hypothalamus. Releasing factor acts on the pituitary by stimulating the release of FSH and LH. Many nerve cells in the brain communicate with the hypothalamus. The messages that they carry and the neurotransmitters that are released by their axons can either stimulate or inhibit the release of GRF from the hypothalamus. This is an example of a phenomenon discussed in Chapter 1: Nervous control systems can influence endocrine (hormonal) control systems. The effects of a nervous control override on reproductive systems will be discussed in greater detail later in this chapter. The complete scheme of reproductive control can be seen diagrammatically in Figure 4-2.

Individual Differences

The cyclic control of the reproductive system is delicate and complicated. As such it is subject to changes brought about by both environmental and hereditary differences. The length of individual cycles, the relative amounts of hormones produced by ovaries and the pituitary, and the sensitivity of target organs to individual hormones are probably all genetically determined and subject to individual variability.

The extent of individual differences among reproductive cycles can be seen in many ways. Some women get pregnant easily and often; for others each successful pregnancy seems almost a miraculous achievement. For some women, menstrual bleeding has occurred just like clockwork ever since menarche. For others, infrequent and/or irregular periods cause physiological or emotional distress.

An example of our individual differences can be seen from results of a study of the length of the menstrual cycles of over 2,000 women during more than 30,000 of their cumulative cycles. In these women the theoretical 28-day cycle occurred only 16 percent of the time. Figure 4-3 illustrates the variability in cycle length for these women, regardless of their age. The proportion of cycles which lasted

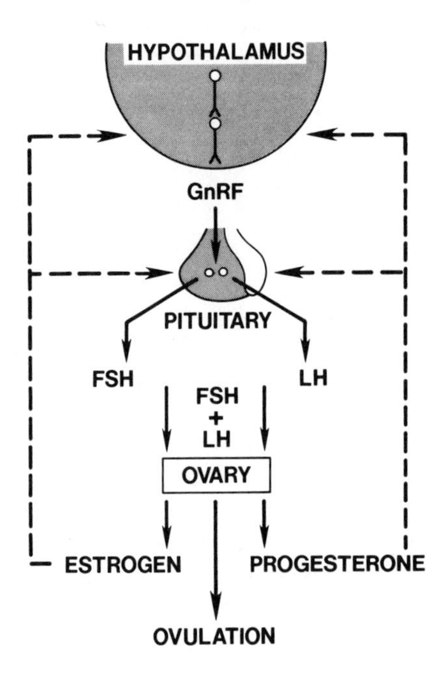

FIGURE 4-2 Hormonal control of cyclic reproductive function: A negative feedback system.

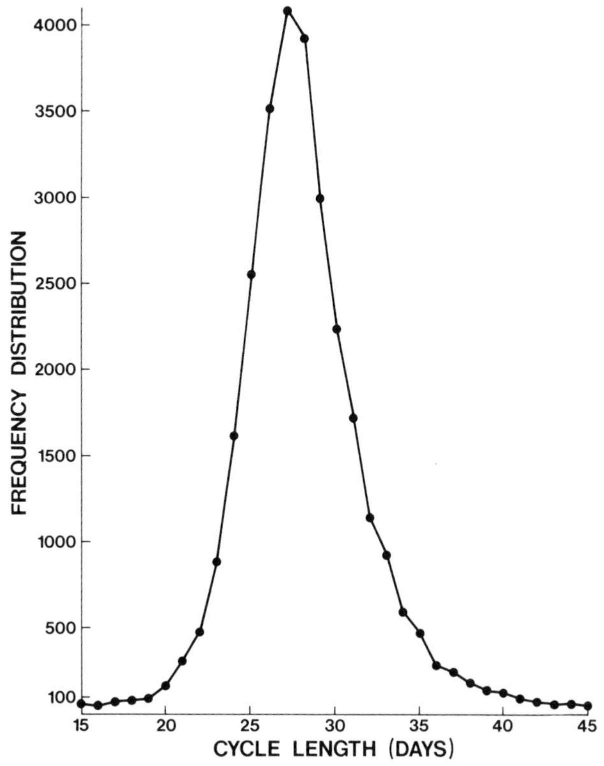

FIGURE 4-3 The length and variability of the human menstrual cycle. From Leonard Chiazze, Jr., Franklin T. Brayer, John J. Macisco, Jr., Margaret P. Parker, and Benedict J. Duffy, "The Length and Variability of the Human Menstrual Cycle," *JAMA* 203, no. 6 (February 5, 1968): 378. Use of these data by permission of Benedict J. Duffy.

between 25 and 31 days was lowest in women younger than 25 and older than 39; it was highest in women who were in the 25-to-39 age group. In this latter group almost 75 percent of the cycles were between 25 and 31 days. This may indicate that the control systems that regulate the reproductive cycle function best during a certain time interval after menarche and before the menopause.

Environmental Influences

Environmental events may also alter the normal function of our cycles. In a study by Matsumoto et al. (1968), a group of Japanese women made records of their cycles during some stress-related conditions, including changes in residence, illness or death of a close relative, and injury. The basal body temperature records of their cycles showed that the women did not ovulate after a stress-related event. The control of the system was probably upset through the hypothalamus. There are other well-documented examples of the effect of some emotional upset on the reproductive cycle. As discussed in Chapter 1, many young women who are away from home for the first time may experience temporary cessation of menstruation. This syndrome is so well known that it has its own name: boarding school amenorrhea. Women in prison camps during World War II reported amenorrhea that was related neither to sexual molestation nor to nutritional deprivation.

In the examples cited above, a common factor is that an external influence is mediated by the hormonal control system to suppress ovulation and/or menstruation. The event itself is insufficient to alter cyclic patterns; it must be the event as it impinges on nervous and hormonal control systems that makes a difference. In addition to having a potential for suppressing ovulation, emotional stimuli may be able to bring about ovulation, so-called coitus-induced ovulation (Clark and Zarrow, 1971). Jöchle (1974) cites as evidence data gathered from interviews with women who became pregnant as a result of rape. The pregnancies occurred on days of the reproductive cycle when, theoretically, pregnancy would not normally occur. Coitus-induced ovulation was proposed as a reason for these pregnancies.

Data about pregnancy resulting from an isolated intercourse during a carefully recorded menstrual cycle have also been gathered (Jöchle, 1974). In these cases pregnancy occurred earlier or later in the cycle than would have been expected. Given the range of cycle length in women (Fig. 4-3) it is also possible that ovulation can occur earlier or later in some women than in others, relative to their overall cycle. It is difficult to document that ovulation can be induced by trauma or by intercourse alone. The current opinion of most reproductive physiologists is that ovulation in human females does not take place without the stimulation from the pituitary hormones, FSH and LH.

Timing of Reproductive Cycles

Some women, like the second subgroup in the coitus-induced ovulation study cited above, keep records of their reproductive cycles. Women are usually more aware of the menstrual bleeding cycle than they are of the occurrence of ovulation during the cycle, so that, when cycle days are counted for whatever reason, it is customary to call the first day of bleeding day 1 of the menstrual cycle. Another way of looking at our cycle is from the point of view of ovulation. Days before ovulation are called day −1, day −2, day −3, . . . , and days after ovulation are termed day +1, day +2, day +3. . . . (Figure 4-6, a representation of the hormone levels during the ovulatory cycle, uses the first method of enumerating cycle days.)

Ovarian and pituitary hormones influence the timing of reproductive cycles. They also affect reproductive structures that change at various stages of the reproductive cycle. The target tissues that are affected by these hormones are the ovaries, the oviducts, the uterus, the vagina, and the breasts.

Ovarian Changes

When we are born, our ovaries contain all the prospective ova (eggs) we will ever have. The numbers of these ova can never be increased, only decreased. Relatively few of them, fewer than 500 out of 500,000, will ever be ovulated. The remainder undergo a poorly understood process called atresia, just as the vast number of embryonic ova degenerate before birth. We do not know exactly which control systems determine the fate of each presumptive ovum, nor what causes the degeneration of so many follicles, the structures which contain the ova. Many of us are postponing our childbearing until we are well-established in our careers. This means that we may be considering pregnancy at a time when our ovaries are past their prime, reproductively speaking. At the same time, we might want to know the effects of oral contraceptive steroids (OCS) on the ovaries particularly when the OCS are used for long periods of time, beginning early in the menarche. Thus far, we do not have adequate information in this regard.

During the reproductive cycle in women who do not take OCS, the ovaries are acted on by the pituitary hormones, FSH and LH. Follicle-stimulating hormone stimulates one of the many nonatretic follicles in the ovary to begin developing. The ovum inside the follicle also undergoes changes that prepare it for fertilization. As the follicle increases in size, the cavity that contains the ovum becomes filled with fluid. This fluid contains increasing amounts of estrogen produced by the follicle cells. When sufficient amounts of LH are produced by the pituitary gland the follicle ruptures and ovulation occurs.

The release of an ovum from the ovary may be sudden and dramatic, or it may take several hours. Animal studies and, more recently, time-lapse photography of the human ovary have shown the ovum emerging from its protective sheath within the ovary. Somehow the wall of the ovary gives way to release the ovum. One theory to explain the mechanisms of ovulation proposes that the LH-receptor complex stimulates cyclic adenosine monophosphate (c-AMP) to produce an enzyme that breaks down the wall of the ovary. The first phase of the ovarian cycle, including the enlargement of the ovarian follicle and its rupture during ovulation is called the follicular phase. Figure 4-4 illustrates the follicular phase of the cycle, and the luteal phase which completes it. The possible role of cyclic AMP is indicated.

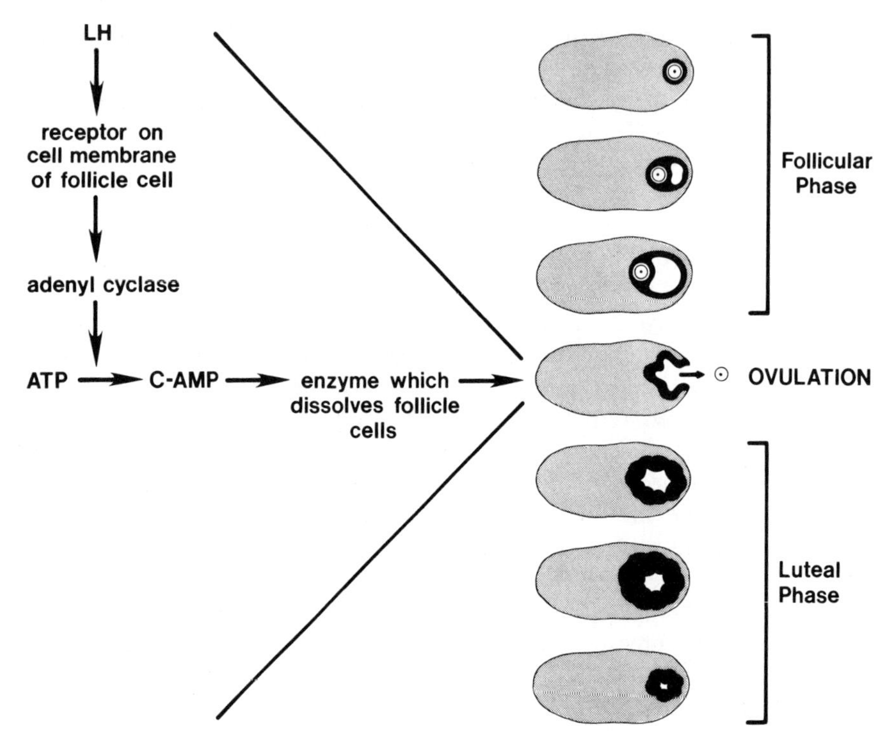

FIGURE 4-4 The ovarian cycle and ovulation. The ovary has been drawn to slightly less than half the dimensions of size of a normal ovary. One theory of ovulation (Espey, 1980) maintains that the enzyme that weakens the follicle cells so that the ovum can be released is a product of LH-induced cyclic AMP.

After ovulation, the follicle cells, under the influence of LH, become organized into a temporary structure, the corpus luteum, which functions like an endocrine gland. During this luteal phase of the ovarian cycle the corpus luteum produces progesterone and estrogen for a short time. If pregnancy does not occur, it degenerates into a small structure that no longer produces hormones. The corpus luteum of pregnancy produces estrogen and progesterone until the placenta takes over this function (see Chapter 5).

Changes in the Oviducts

Other structures in the reproductive system are influenced by cyclical hormone production. The oviducts, passageways between the ovary and uterus, are lined with cells that have cilia. The cilia are hair-like extensions of the cells that move like long grass blown by the wind. When the cilia are stimulated by estrogen they beat rhythmically. This rhythmic beating can move sperm up to oviducts toward the ovum, or it can move the ovum down toward the uterus. Ovarian hormones also stimulate smooth muscles in the oviduct. The contraction of the muscles augments the effect of the cilia. The movement of the ovum down the oviduct depends upon just the right concentration of estrogen. If the level of estrogen is too high, the ovum is moved too quickly to the uterus, and the endometrium is not ready for the ovum when it arrives.

Uterine Changes

At the beginning of the ovulatory cycle the lining of the uterus, the endometrium, is stimulated by estrogen to grow and increase in depth. During the first phase, called the proliferative phase, blood vessels and glands develop in the endometrium. After ovulation, progesterone completes the preparation of the endometrium for the possible arrival of a fertilized ovum. During this secretory phase, the glands of the endometrium produce a nutritive substance that would be the first food for a developing embryo. A trained observer can determine the stages of the ovulatory/menstrual cycle by examining a small piece of endometrial tissue. This examination, an endometrial biopsy, is often done for other purposes, for example, to determine if cancer has invaded uterine tissue. Unless the endometrium is sustained by the continued hormone production of pregnancy the tissue regresses. Blood and mucus are sloughed off and pass out of the body through the vagina. This is the familiar menstrual bleeding. Figure 4-5 shows what happens to the endometrium during the preovulatory, proliferative phase and the postovulatory, secretory phase of the cycle.

Vaginal Changes

The cervical mucus, which can be found in the vagina, has different consistencies, each of which is characteristic of a certain stage of the ovulatory cycle. This change in the cervical mucus is influenced by the ovarian hormones. Estrogen causes the mucus to become thicker; progesterone causes it to thin out. At about the time of ovulation, when estrogen levels peak, the mucus is thickest; during the postovulatory phase of the cycle it gradually thins out as progesterone is secreted. Some women can be and have been trained to observe changes in their cervical mucus as an aid to determining the time in the cycle when they are likely to be ovulating.

Changes in the vaginal cells during the reproductive cycle were studied by Dr. Papanicolaou, before he developed his method for cancer detection (the Pap smear). Estrogen and progesterone stimulate characteristic changes in the vaginal cells. These cells from the vaginal epithelium are exfoliated (shed) regularly. Vaginal cells

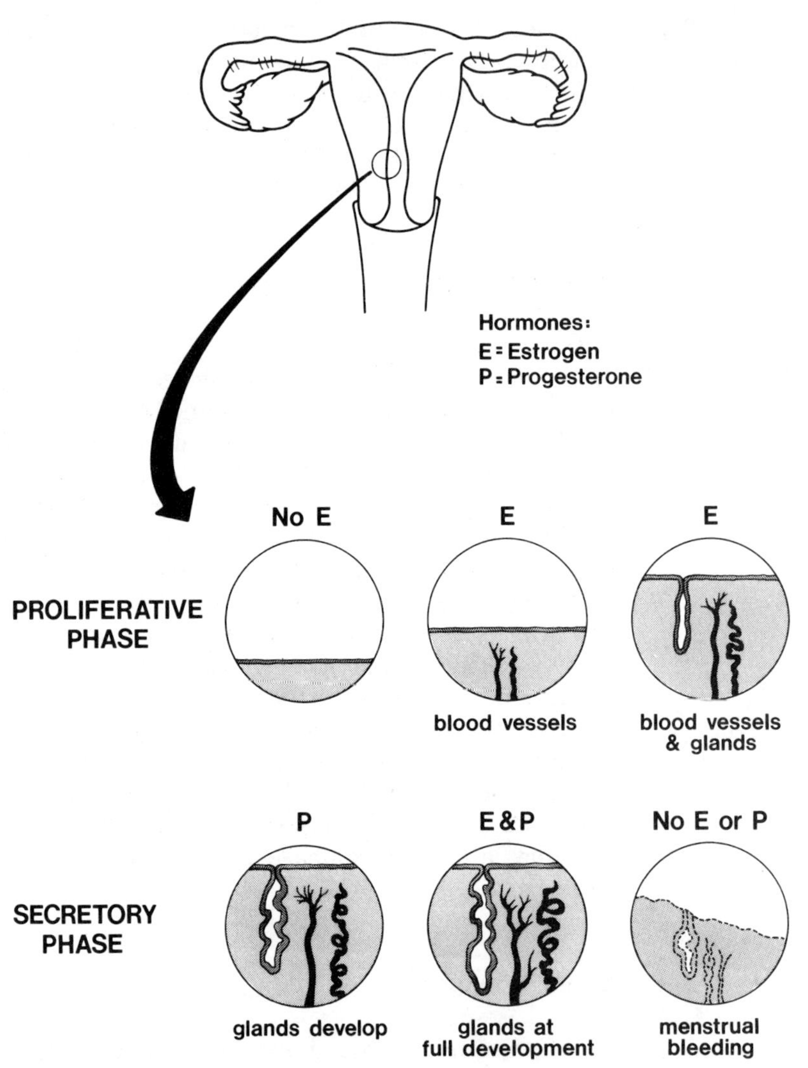

FIGURE 4-5 Changes in the lining of the uterus (endometrium) during the menstrual cycle.

can be transferred easily from the epithelium to a microscope slide by using a cotton-tipped swab or a specially designed pipette. After the cells have been processed and stained, their nuclei show changes characteristic of the hormones that have stimulated them. Vaginal cytology, the study of cells from the vagina, was formerly used to determine a woman's reproductive status: pubertal, reproductive, pregnant, or menopausal. If a woman were reproductive, this technique could show characteristic differences in the cells during each phase of the cycle. Since it is now possible to determine the stages of the reproductive cycle by measuring estrogen and progesterone levels in the blood, vaginal cytology is no longer used for this purpose.

However, examination of the cells from the vagina and the cervix as in the Pap smear has been extremely valuable in detecting early stages of cervical cancer (Chapter 8).

Other Cyclical Patterns

We have seen in Chapter 3 that breast development during puberty depends upon the secretion of ovarian hormones. After menarche there are cyclic changes in breast size that may or may not be noticed. The changes in breast size that are caused by hormonal influence are augmented during pregnancy when the breasts are being prepared to nourish the developing infant. This will be discussed in greater detail in Chapter 5.

Current methods of hormone measurement show that there is

FIGURE 4-6 Hormonal fluctuations during the menstrual cycle. From Leon Speroff and Raymond L. Vande Wiele, "Regulation of the Human Menstrual Cycle," *Am. J. Obstet. Gynecol.* 109, no. 2 (1971) 234–47, by permission of C. V. Mosby Company and the authors.

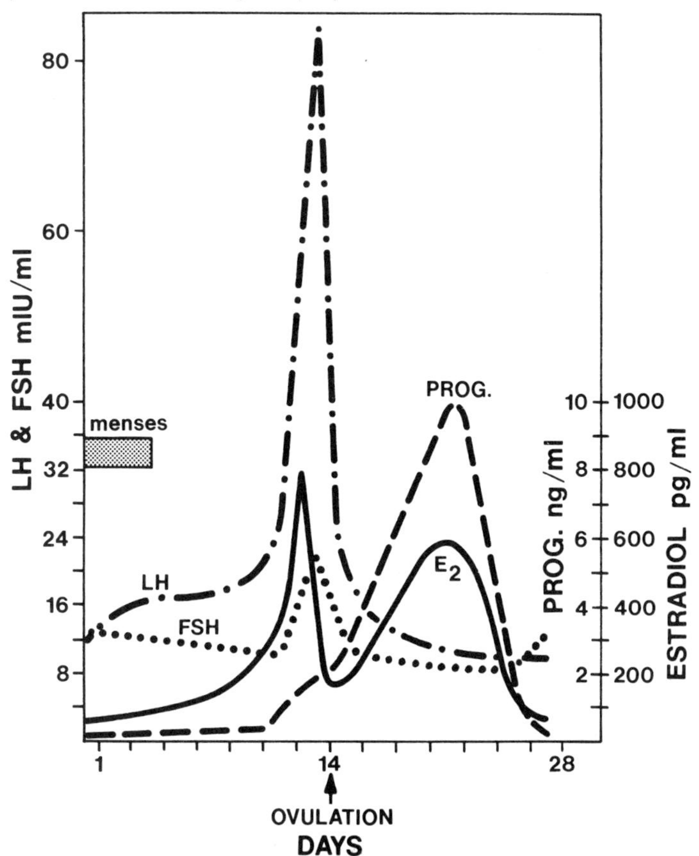

a pattern in the production and release of the hormones that control our reproductive cycles. The rhythm of ovulatory cycles depends upon critical concentrations of pituitary and ovarian hormones. The concentration of these hormones, measured in terms of units of hormone per cubic centimeter or per milliliter of blood plasma, varies during the ovulatory cycle. An example of the measured levels of estrogen, progesterone, LH, and FSH in the plasma of a woman during her reproductively fertile years appears in Figure 4-6. Individual differences in the relative amounts of these hormones and in the way these amounts change over time are probably due to differences in our gene-directed hormone production. Some other causes of the alteration of these patterns have been discussed earlier in the chapter. Pregnancy and menopause also bring about changes in the monthly cycle.

The control systems that determine our cyclic fertility and sterility differ from those of our male partners in several important ways.

REPRODUCTIVE MECHANISMS IN OUR MALE PARTNER

Control Mechanisms

Puberty in men is initiated by control mechanisms that operate, as ours do, through the hypothalamus, pituitary, and gonads (Fig. 4-7). These mechanisms include a releasing factor from the hypothalamus, the pituitary hormones LH and FSH, and testosterone from the testes.

The male reproductive control system differs from ours in that it is continuous, not cyclic. As we have seen (Fig. 4-6), we have

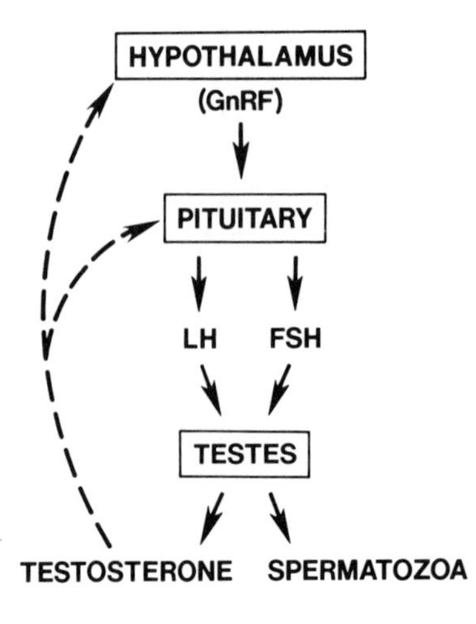

FIGURE 4-7 Hormonal control of male reproductive function: A negative feedback system.

a mid-cycle peak in LH and FSH, which stimulates ovulation, the production of one (usually) ovum. In contrast, the pituitary gland of a normal male continuously produces, from puberty to old age, enough LH and FSH to assure the ongoing production of both testosterone and sperm cells in the testes.

PRODUCTION OF SPERMATOZOA The male testes, unlike our ovaries, are located outside the body in the scrotal sacs. Their location is important because a man's ability to produce normal spermatozoa is related to the temperature of the testes. The temperature must be *below* the body temperature. Experiments in men and in laboratory animals have shown that testes exposed to higher temperatures cannot produce normal sperm (Robinson and Rock, 1967).

The cells in the testes that produce spermatozoa are constantly renewed through cell division. The sperm-producing cells in the testes function throughout a male's postpubertal life, in contrast to our reproductive cells which are never replenished and which gradually diminish in number until our reproductive lives end at the menopause.

The formation and release of one ovum takes about a month. Formation of mature spermatozoa (sperm cells) takes over two months. Since sperm formation is going on in both testes continuously, millions of mature sperm are generally ready to be ejaculated at any one time. The long period of time and the many millions of cells involved in sperm production may partially explain why we have, at present, no male oral contraceptive pill. If this long sperm-producing cycle were to be interrupted arbitrarily at any one point, abnormal sperm could be produced. These sperm might contribute to the development of an abnormal child.

TESTOSTERONE PRODUCTION Testosterone is produced by the testes, the adrenal gland, and, in small amounts, by the ovaries. The larger amounts of testosterone produced by the testes are responsible for the physical characteristics we associate with maleness—muscle development, voice depth, and the development of facial hair. Testosterone is also necessary for the continuation of sperm production and for the normal functioning of the male genital duct system.

The effects of testosterone can be seen by examining cases of testosterone deprivation or administration. Eunuchs, males who have been castrated (had their testes removed) before puberty have higher-pitched voices and more rounded body contours than those of normal males. On the other hand, supplemental testosterone administered to males already producing normal amounts of this hormone seems to increase the sex drive. Behavior patterns that were present before testosterone administration are not redirected; for example, if homosexuals are given supplemental testosterone their sex drive toward males is increased (Ganong, 1975).

SPERM SURVIVAL Semen is ejaculated from the penis into the vagina during heterosexual intercourse. Semen contains sperm in a liquid produced by the seminal vesicles, Cowper's glands and the prostate gland. The function of the liquid is to provide a medium for optimum sperm survival. Because nutrition and pH regulation are furnished in this liquid medium, sperm cells can and do live for many weeks in the male genital tract. Once they have been ejaculated in the semen their maximum life span is 24 to 72 hours at body temperature.

The survival of ejaculated sperm is a critical factor in conception. This survival time has been prolonged artificially by the use of sperm banks, where sperm are stored frozen at low temperatures. Stored sperm can be used for artificial insemination. Not all men produce spermatozoa that are capable of fertilizing an ovum. Abnormally shaped sperm, sperm that cannot move vigorously, or sperm of low concentration in the semen may not be able to fertilize ova.

Male Sexual Response

Penis erection and the ejaculation of semen are reflex nervous control systems. They are integrated at the spinal cord, but they are subject to the influence of the cerebral cortex. Sexual response in the male may be initiated by visual imagery or by direct stimulation of erogenous zones. Receptors in the penis are activated and the message is carried to the spinal cord. Inputs from the brain can either inhibit or enhance the nervous system message to effectors in the smooth muscle. Small arteries in the penis dilate, causing expansion of erectile tissue in the penis. Figure 4-8 illustrates the nervous control system for erection and ejaculation

Ejaculation of the semen is also a spinal reflex with similar nerve pathways. Increased stimulation of the receptors brings about contraction of smooth muscles in the genital ducts and of skeletal muscle at the base of the penis. Orgasm in the male is the result of the rhythmic contractions of the penis and genital ducts during ejaculation.

Sexual responses in the male were investigated in the laboratory by Masters and Johnson (1965). They divided the response into phases, which have characteristic physiological responses:

PHASE	*RESPONSE*
1. Excitation (follows stimulation)	*Erection.* *Urethra lengthens.* *Testes elevate in scrotum.*

2. Plateau	*Penis becomes hard.* *Testes increase in size.*
3. Orgasm	*Contraction of skeletal and smooth muscle.* *Ejaculation of semen.* *Increases in pulse rate, blood pressure, respiration.*
4. Resolution	*Loss of erection.* *Testes return to normal size and position.* *Psychological and muscular relaxation.* *Refractory period—no erection and ejaculation possible. The time of this refractory period varies among individuals.*

FIGURE 4-8 Sensory and motor nerve pathways in erection and ejaculation.

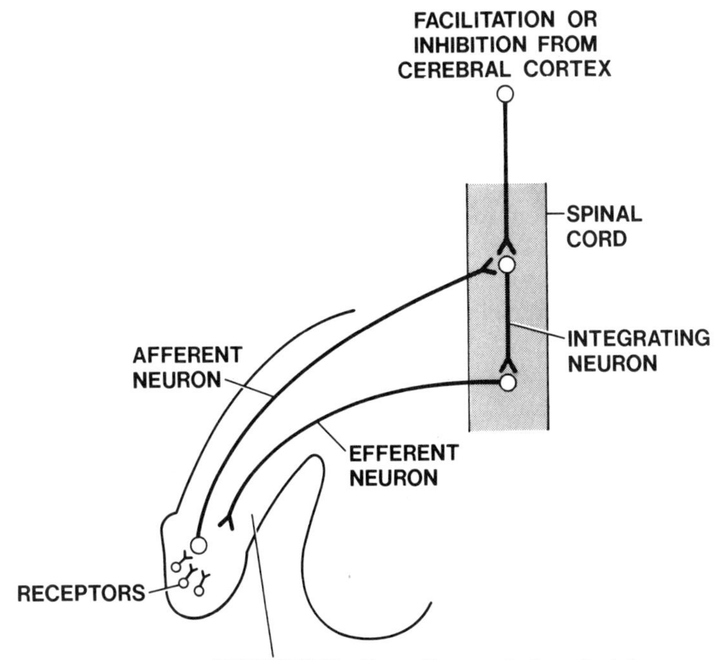

Stress may affect testicular function in the male, so that he cannot produce normal sperm in adequate numbers. The basic reflex mechanisms that control erection and ejaculation may also be altered by any number of psychosocial influences as well as by physiological disorders. Alcohol is a central nervous system depressant, and its excessive consumption may make a man temporarily impotent. Impotence in the male is the inability to respond to erogenous stimulation by erection and ejaculation. It is a separate problem from sterility, which refers to the production of sperm, their number, shape, and movement. In theory, a fertile and potent male can impregnate his partner unless some contraceptive measures are used.

The Male and Contraception

The various methods of contraception and their relative effectiveness will be discussed in detail in Chapter 6. Here we shall look briefly at the contraceptive methods involving male responsibility and control.

COITUS INTERRUPTUS Coitus interruptus or withdrawal is an often-used and least effective technique of fertility regulation. Intercourse stops with the removal of the penis from the vagina prior to ejaculation. The reflex ejaculatory mechanisms must be shut off by voluntary intervention, and the importance of timing creates physiological as well as psychological problems. Moreover, a few drops of semen may leave the penis when it is erect, before it is even inserted in the vagina, and this semen may contain enough sperm to cause pregnancy. Most couples who have alternatives will probably use other methods of contraception.

THE CONDOM The condom has been used for many years, both as a contraceptive and for protection against venereal diseases. The modern condom, if properly used, is very effective as a contraceptive. In three recent British studies, condoms had a failure rate of 1 per 10,000 uses. This means that, when the condom is used correctly, during *every* coital activity, only 1 coital act per 10,000 will result in pregnancy. Condoms also are effective against the transmission of most venereal diseases. In a study made during the Vietnam war, 27 percent of the soldiers who used no method of protection contracted a venereal disease, while none of the soldiers who always used condoms developed infection. A recent article in *Consumer Reports* (1979) tabulated the advantages and disadvantages of condom use, as expressed by polled readers (about 1,000 men and 800 women). The results are shown in Figures 4-9 and 4-10.

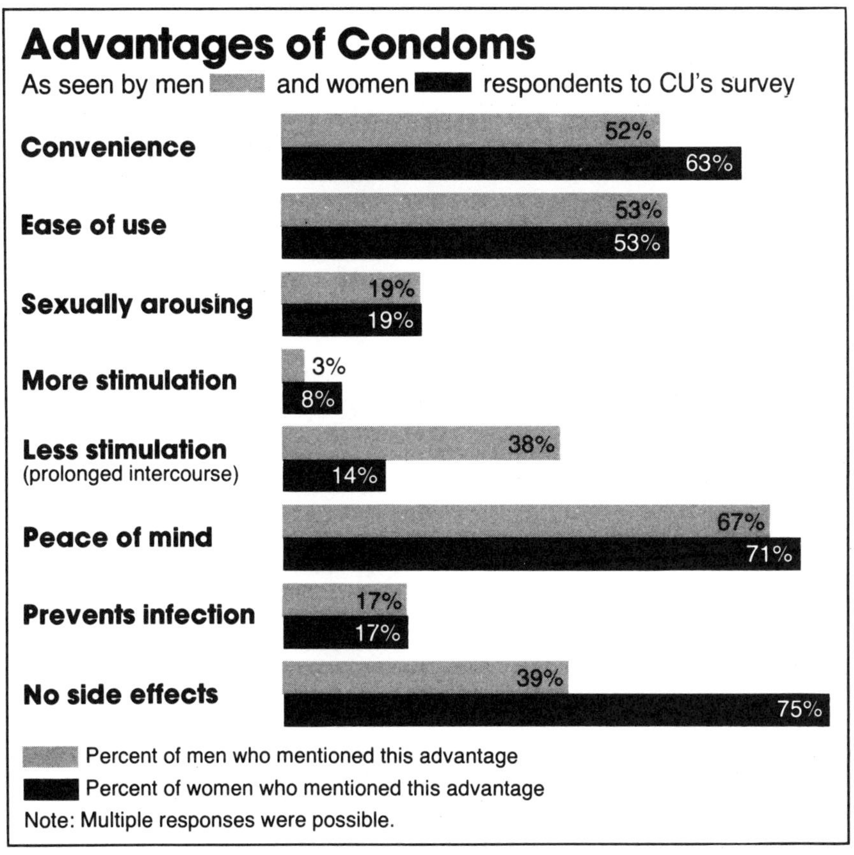

FIGURE 4-9 Copyright 1979 by Consumers Union of United States, Inc., Mount Vernon, NY 10550. Excerpted by permission from *Consumer Reports,* (October 1979).

VASECTOMY Sterilization of the male, called vasectomy, involves a cut in the vas deferens, the duct that delivers sperm to the urethra. Both sperm and testosterone continue to be produced in the testes. Only the *delivery* of sperm during intercourse is affected. The semen deposited in the vagina contains no sperm. In recent years, this simple operation, which can be done as an office procedure, has become increasingly popular. At the present time, no persistently harmful side effects are known. If a man who has had a vasectomy changes his mind, it is possible, in some cases, to rejoin the cut ends of the ducts. The man, however, may still be unable to father children. Some men form antibodies to their own sperm after a vasectomy, and these sperm antibodies will prevent sperm from fertilizing the ovum. If our partners are considering vasectomy as a method of fertility control, we should understand that there is no certainty, at this time, that such a decision can be reversed after the operation.

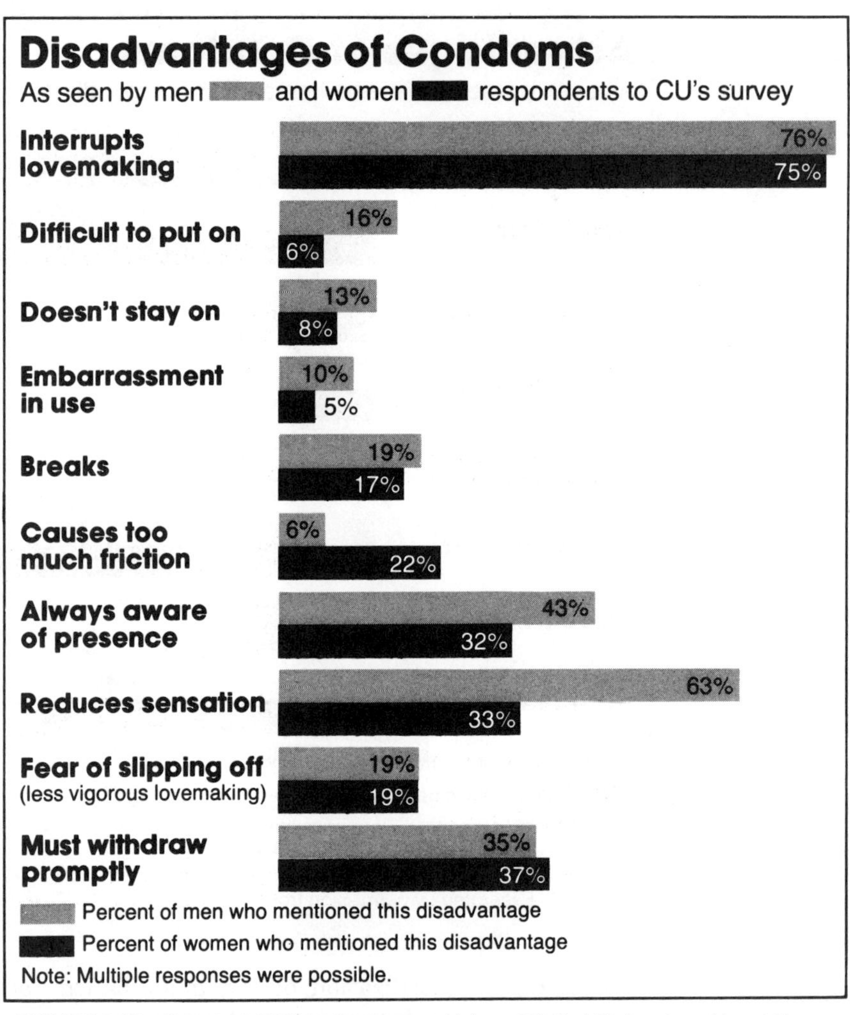

FIGURE 4-10 Copyright 1979 by Consumers Union of United States, Inc., Mount Vernon, NY 10550. Excerpted by permission from *Consumer Reports,* (October 1979).

A MALE PILL? It is easier to interrupt a once-a-month event than it is to stop the continuous production of millions of spermatozoa. This does not fully explain why the development of chemical contraception for the male has lagged behind similar technological progress made in our behalf. Many men are willing to assume responsibility for the control of their partner's fertility.

Some current attempts to develop a male contraceptive pill involve the inhibition of GRF release from the hypothalamus and subsequent interference with testosterone as well as sperm production (Fig. 4-7). Since testosterone is necessary for the maintenance of male secondary sex characteristics, most men would find this an undesirable side effect. Another chemical contraceptive tested in males had a different side effect: intolerance of alcohol. Other compounds that interfere with sperm production are being tested, but none is available for use at the present time.

MECHANISM OF THE FEMALE SEXUAL RESPONSE

"Intercourse," "having sex," "coitus," or whatever we may call it means different things to different people. The psychology of a woman's sexual activity is deeply personal. The physiology of female sexual activity, like that of the male, can be described objectively, in terms of what happens in our bodies when we are sexually stimulated (Masters and Johnson, 1965).

PHASE	*RESPONSE*
excitation (somatic or psychic stimulation)	*lubricating fluid in vagina.*
	clitoral swelling.
	enlargement and erection of nipples.
	lengthening and expansion of vagina.
plateau	*vasocongestion in labia, vagina, uterus.*
	muscle tension.
orgasm	*vascongestion and muscle tension.*
	contraction of smooth muscles, vagina and uterus.
	increased pulse rate, respiratory rate, blood pressure.
resolution	*muscular and psychological relaxation.*
	return to normal size: clitoris, vagina, uterus.
	no rest required before another orgasm.

The spinal reflex that controls orgasmic responses in our bodies are similar to those in the male. Receptors are located in the nipples and clitoris. The same kind of override of the spinal reflex can be brought about by inputs from the brain ("I shouldn't," "I can't," "Nice girls don't," "The baby is crying").

Why does he "come" when we "stay"? The man's orgasm seems to be inevitably connected with his ejaculation. When his ejaculation is over, we may be left wondering why anyone thinks sex is all that great.

We need more time, time during foreplay, time for stimulation when the penis is in the vagina (intromission). A reliable source of information about human behavior (American Medical Association, 1972) enforces our experience that foreplay is very important to us. If stimulation of the receptors in the nipples and clitoris lasts from 1 to 10 minutes, about 40 percent of us will experience an orgasmic response. This response percentage increases to 90 if stimulating foreplay lasts for 20 minutes. The clitoris is stimulated during intromission, but few if any women have a climax if this time is less than 1 minute. With an intromission time of 1 to 11 minutes, 50 percent of women experience orgasm, and almost all achieve it after 16 minutes of stimulation during coitus. According to the Kinsey report, three-fourths of males reach orgasm and ejaculate in less than 2 minutes, many in 10 to 20 seconds.

What we seem to have is a problem of timing. Most of us need longer periods of time than our male partners for stimulation of the receptors that initiate sexual response (orgasm). Our understanding of this need is important. We may have experienced the need without understanding its basis. Communication of this need into patterns of love-making may help us experience why sex is all that great.

Women who remain in good health may continue their sexual activity well into old age. Meanwhile, however, there comes a time when their reproductive ability ceases.

THE MENOPAUSE

The Onset of Menopause

At the opposite end of the reproductive continuum from menarche (Fig. 4-1), the menopause marks the end of a woman's ability to bear children. During puberty the ovaries gradually begin their cyclical functioning to produce ova, estrogen, and progesterone. During the climacteric, the period of time leading up to the menopause, the ovaries gradually cease producing gametes and hormones.

The onset of vaginal bleeding was the sign that our bodies were approaching sexual maturity, and our monthly bleeding periods let us know that we were still producing hormones and probably ova. As menopause approaches, however, we detect changes in our bleeding patterns. These changes are not the same for every woman. In some, the bleeding may be excessive and prolonged for several months before the menopause. In others, the amount of blood loss and the length of flow may be gradually reduced during the one to two years before the menopause.

TABLE 4-1 Probability of Menopause According to Duration of First Amenorrhea and Age

First Amenorrheal Interval (Days)	*Age* 45–49	50–52	*Over 53*
60–89	6.0	21.6	35.2
90–119	12.8	30.4	47.4
120–149	25.1	42.4	56.2
150–179	39.5	56.4	65.6
180–209	45.5	65.2	71.9
210–239	55.3	73.1	78.4
240–269	63.5	81.5	85.3
270–299	74.1	86.3	88.6
300–329	83.0	90.3	90.5
330–359	88.2	92.5	92.9
360 +	89.5	93.6	95.5

Source: R.B. Wallace, B.M. Sherman, J.A. Bean, A.E. Treloar, and L. Schlabaugh, "Probability of Menopause with Increasing Duration of Amenorrhea in Middle-Aged Women," *Am. J. Obstet. Gynecol.* 135, no. 8 (December 15, 1979): 1022, by permission of C. V. Mosby Company and the authors.

The onset of irregular menstrual cycles in a woman who is more than 45 years old is a recognized clinical indication of impending menopause. When does a given interval of amenorrhea signify that the menopause, the permanent cessation of menses, has occurred?

The probability of natural menopause associated with various intervals without menses has been calculated (Wallace et al., 1979). The investigators gathered data from a prospective, that is, ongoing, study of women approaching the age of menopause. These women had been participants in a study of menstrual and reproductive events beginning in 1961. They had recorded their menstrual cycle lengths during that time. None of the women had taken oral contraceptive steroids. The probability of their entering the menopause after different amenorrheal intervals is summarized in Table 4-1.

Thus, the transition from the regular menstrual intervals which we experience during the fertile part of our reproductive lives to the final cessation of menstruation is usually a gradual one. During this transition women may experience a phase of increasing mean cycle length and increasingly irregular cycle intervals. In some women, prolonged intervals may be interspersed with extremely short ones. Physiologically, these variations represent irregular development or incomplete maturation of the ovarian follicles. The aging ovary is the focus of changes that characterize the menopause.

The Aging Ovary and Its Hormones

Chronological aging brings about continuous and irreversible changes in the structure and functioning of our bodies as a whole. Some scientists believe that aging itself may be studied in terms of imbalances in control systems, the nervous and endocrine systems, which regulate homeostasis (Timiras, 1979). The aging ovary, in this respect, might serve as a model by which to study aging and senescence, the detrimental changes that accompany aging.

The ovary begins to age before we are born. We have seen that atresia, or follicular degeneration, commences during intrauterine life and continues thereafter. By the time of the menopause, something has happened to the ovary and its nonrenewable supply of remaining prospective ova. It does not respond to stimulation by LH and FSH, and so it does not produce ova or secrete hormones. Gonadotropin levels continue to rise because there is no negative feedback by estrogen and progesterone (Fig. 4-11). The hormonal condition of the menopausal woman and her inability to conceive a child are brought about by the age-related changes in her ovaries.

Although many hormone-mediated functions are altered during aging, the nature and mechanisms of these changes have yet to be fully explained by medical science. A number of these changes have been related to changes in the concentration of the receptors in the hormone-responsive cell (Roth, 1979). Another explanation for aging in the ovary might be sought in the mechanisms that control the release of GRF from the hypothalamus or of LH and FSH from the pituitary gland.

Whatever their regulatory mechanisms, the pattern of hormone secretion for menopausal women—decreased estrogen with increased LH and FSH—is the basis for menopausal changes in our bodies. We no longer produce the ova or the hormones we produced during our reproductive years. At this stage in our lives (45 and over) few of us are concerned about becoming mothers, again or for the first time. We may, however, miss the hormones.

Distinguishing the phenomena of menopause from those of aging is difficult because menopause and aging overlap. Some changes that occur before, during, or after the menopause may be a result of aging per se while others are known to be directly related to the lack of estrogen. The reproductive structures in our bodies that develop during puberty under estrogen stimulation are the ones that may atrophy when estrogen stimulation is withdrawn.

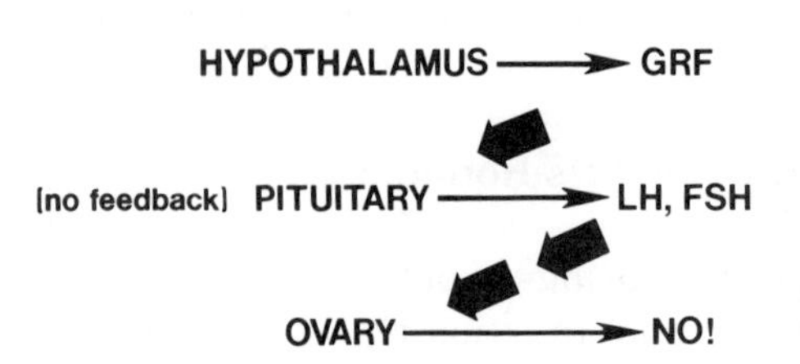

FIGURE 4-11 The unresponsive ovary of the menopause.

Reproductive System Changes during Aging and the Menopause

After the age of 40 the frequency of ovulation usually decreases, possibly because the follicles become less sensitive, with time, to circulating gonadotropins. The inability to develop new ova and the continuing depletion of ova throughout life can be considered as an aging process, related to, but not caused by, the menopause.

With the decrease in the number of follicles, total gonadal hormone secretion also decreases. The irregular production of these hormones, especially of estrogen, is the source of irregularities in menstrual patterns. The transition between hormone production during reproductive cycles and the menopause can be seen in hormone levels during the premenstrual period. Figure 4-12 illustrates several menstrual cycles in a woman who was entering the menopause. Cycles I and II show the high levels of FSH and LH secretion, with corresponding low levels of estrogen and no measurable amounts of progesterone. In cycles III and IV the lower LH levels indicate that some feedback from the ovary might be taking place. Cycle III shows some detectable progesterone production, which indicates that ovulation probably occurred during this cycle. Pregnancy is a possibility in this premenopausal period when the hormone secretion patterns have not yet stabilized into those characteristic of the menopause, as in cycles I and II.

The ovary is one of the first organ systems to reveal significant aging, a process that is not limited to the climacteric. The aging process in the ovary began before birth and continued throughout reproductive life. With the end of reproductive capability at the menopause, steroid hormone production is greatly decreased, but does not cease altogether. Some estrogen production occurs after the menopause. A male hormone is produced by the adrenal gland and by the nonfollicular part of the ovary. This hormone, androstenedione, may be converted into weak estrogens in tissues such as fat, liver, and kidney. Obese persons tend to produce more estrogen by this method (Schiff and Wilson, 1978) since fatty tissues are sources of estrogen in postmenopausal women. This is a mixed benefit because obesity may be a predisposing condition for breast cancer (Sherman et al., 1979) and endometrial cancer (Siiteri, 1979).

Both aging and decreased estrogen supply cause atrophy of the uterus. The endometrium, or lining of the uterus, is one of the target organs for the ovarian hormones. Since there is considerable variation in the secretion of these hormones by the ovary, different bleeding patterns during the climacteric can result. During the postmenopausal period the conversion of ovarian and adrenal androgens may be high enough to cause endometrial proliferation and even postmenopausal bleeding in some women.

The cells that are shed (exfoliated) from the vagina can indicate

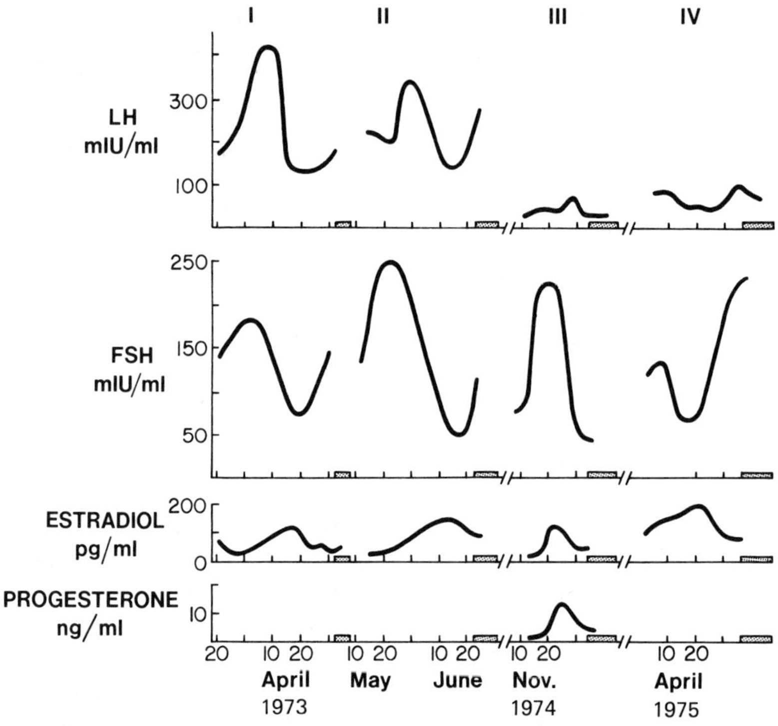

FIGURE 4-12 Premenopausal changes in hormone levels. From B. M. Sherman, J. H. West, and S. G. Korenman, "The Menopausal Transition: Analysis of LH and FSH, Estradiol and Progesterone Concentrations during Menstrual Cycles of Older Women," *J. Clin. Endocrinol. Metab.* 42 (1976): 629. Adapted by permission of J. B. Lippincott Company and the authors.

hormone levels. The cell types that appear in vaginal secretions from women in their reproductive years differ from the cell types that can be seen in the vaginal fluid of menopausal women. Decreased estrogen is the cause of vaginal cytology changes and also of decreased levels of glycogen in the vaginal fluid. Because fewer normal microorganisms such as lactobacilli can grow when the glycogen is decreased, more contaminating bacteria such as streptococci and staphylococci take over the vaginal environment. The growth of these microorganisms can lead to vaginal discharge and infection. They are correctable by estrogen replacement therapy (Schiff and Wilson, 1978).

The vaginal epithelium, which is normally sustained by estrogen, becomes thin and fragile in its absence. Some scarring and shortening of the vagina may occur during the menopause. The condition of the vagina may be a source of pain during intercourse.

Do menopausal and postmenopausal women enjoy sex? The answer to this question is "Yes, if. . . ." The clitoral response is normal in older women, and they can achieve orgasm through clitoral stimulation. The vagina, however, loses its ability to undergo expansion.

The thinning and drying of the vaginal mucosa may contribute to varying degrees of pain during intercourse. Before penile-vaginal intercourse, older women may require longer periods of stimulation to enable the vagina to become well lubricated. The older woman may also need some form of estrogen replacement therapy to correct the menopausal condition of her vaginal epithelium.

Older women are capable of having an active sex life. The more frequent their sexual activity, the better their capacity for sexual response. If coitus does not occur at all during the menopausal years, the vagina may shrink and become a small, semirigid tube that is no longer distensible.

Vasomotor Disturbances (Hot Flashes)

Nearly 75 percent of women approaching menopause experience some form of vasomotor instability (Perlmutter, 1978). The term "vasomotor instability" refers to changes in the diameter of blood vessels in the skin. These blood vessels contract (decrease in diameter) or dilate (increase in diameter) because of nerve impulses from the autonomic nervous system. Since this part of the nervous system is not regulated by impulses from the brain, there is no voluntary control of the blood vessel changes that precipitate the hot flash.

A hot flash has been described as a sudden feeling of warmth in the upper part of the body or all over the body. The frequency with which this symptom occurs is highly variable and the flash can last from less than a minute to almost an hour. In a study of 929 hot flashes in 67 different women, about 49 percent were mild (barely noticeable, no perspiration, lasted less than one minute, and did not interfere with activity). Approximately 39 percent were moderate (a warm feeling more noticeable, perspiration for two to three minutes, removal of a layer of clothing). The remainder, about 12 percent, experienced severe hot flashes which involved intense heat and profuse perspiration. These severe hot flashes lasted longer than three minutes and provoked feelings of distress and inability to cope (Voda, 1979).

At the present time no one knows why women experience these symptoms prior to and during the menopause. Menopausal hot flashes were studied physiologically by Molnar in 1975. He measured internal and external temperature, electrocardiogram (ECG), heart rate, and sweating in a 59-year-old menopausal subject. He noted that the onset of the flash could be measured by a rapid heartbeat, changes in the ECG base line, and detectable sweating. A fall in the skin temperature occurred when the blood vessels dilated. This followed a rapid warming that was first detectable in the fingers and toes. Molnar suggested that the hot flash might be caused by a sudden discharge of a neurotransmitter substance that, in turn, was induced by gonadotropin stimulation.

The measurement of temperature changes in the fingers has been used in subsequent studies of menopausal hot flashes. A recent study (Meldrum et al., 1980) used this measurement combined with newer methods of measuring small changes in hormones. Six women were studied who were experiencing frequent and severe hot flashes after menopause. They were not receiving estrogen therapy. During the eight-hour period they were being studied, their finger temperature was being recorded. At the same time, small samples of blood were withdrawn every 15 minutes from an intravenous catheter. These samples were analyzed for hormone changes. Changes in finger temperature and in LH and FSH levels were recorded with reference to the subjective report of a hot flash feeling. There was a close relationship in time between the releases of LH and the hot flashes. Meldrum and his associates attribute these releases of LH, in part or wholly, to increases in the release of GRF into the circulatory system. It is possible that either the release of GRF or of the neurotransmitters that stimulate its release are somehow connected with the hot flash experience. The neurons that produce GRF in the hypothalamus are anatomically close to the thermoregulatory centers in this area of the brain (Reaves and Hayward, 1979). At the present time there is no direct evidence of a cause-and-effect relationship between LH release and hot flashes, and we do not know why hot flashes are so severe in some women and relatively mild in others. Further research into the vasomotor control systems of menopausal women may identify the source of this symptom of the menopause.

Osteoporosis

Although hot flashes, which affect most women during the menopause, are uncomfortable, they are rarely debilitating. Osteoporosis is. Osteoporosis is a condition characterized by fragile bones. These bones fracture easily under stresses that would not affect normal bones. Osteoporosis is a significant problem for older women. About 25 percent of all white women have had one or more fractures by the time they are 65. Men, as well as women, begin losing bone mineral around age 40, but women, who have lighter bones to start with, may run a higher risk of fracture. Women's bone losses accelerate at the menopause, during which time they may lose an average of 1 percent of their bone mass per year. Age and the menopause combine to upset the homeostatic control systems that regulate bone formation and resorption. What are these regulatory mechanisms?

The dead bones of human skeletons that are studied in anatomy laboratories do nothing to convey the dynamism that characterizes bone as a living tissue. Homeostasis, that process by which the body changes constantly in order to maintain itself, is illustrated by the mechanisms that control the resorption and release of calcium from storage sites in the bone.

Bone tissues are composed of bone cells called osteocytes, which are illustrated in Figure 4-13. These cells are capable of using calcium to form the substance of bone. Other bone cells, called osteoblasts (Fig. 4-13), are less differentiated. They are located in the periosteum, the tissue around the bone. Osteoblasts may develop into osteocytes or into osteoclasts, which break down bone to release the calcium. The dynamic equilibrium between bone formation (incorporation of calcium into bone) and bone resorption (release of calcium from bone) depends on the relative activity of the two cell types, osteocytes and osteoclasts, respectively.

Bone formation and resorption are mechanisms for storing and releasing calcium. Calcium is essential for many metabolic functions including muscle contraction. Two control mechanisms act to regulate blood calcium levels. The first responds to a drop in plasma calcium levels by stimulating the parathyroid hormone (PTH), which has several metabolic functions:

1. *PTH stimulates the osteoclasts to break down (resorb) bone. This releases calcium into the circulation.*
2. *PTH acts on the cells of the kidney tubule to stimulate increased calcium reabsorption.*
3. *With vitamin D, PTH stimulates increased calcium absorption in the intestine.*

The net effect is to raise levels of blood calcium.

The second responds to increases in blood calcium by stimulating the operation of another negative feedback system which has re-

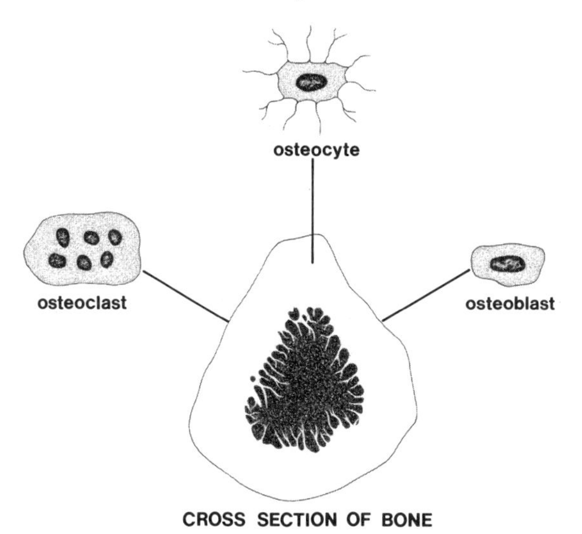

FIGURE 4-13 Cell types in bone formation and resorption.

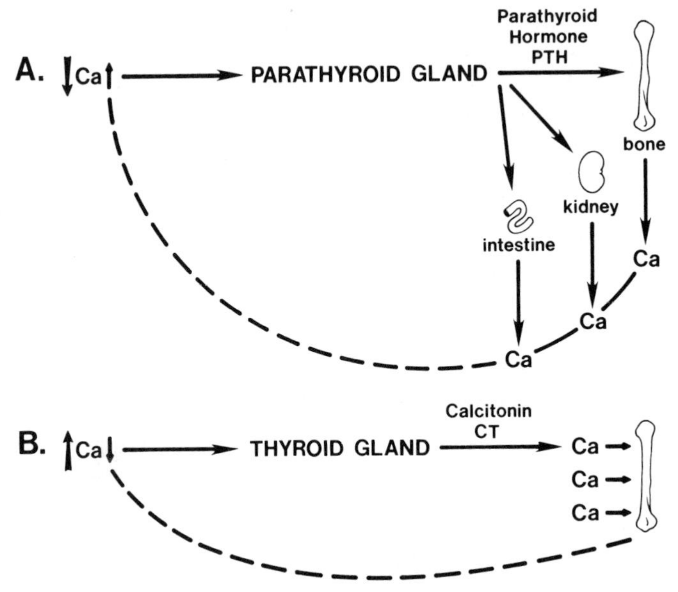

FIGURE 4-14 Negative feedback systems in calcium homeostasis.

ceptors in the thyroid gland. When calcium levels are increased beyond the set point, the thyroid gland releases calcitonin (CT). This hormone stimulates the activity of the osteoblasts. They develop into active, calcium-using cells or osteocytes. This formation of new bone tissue decreases the blood plasma level of calcium. Figure 4-14 shows the operation of PTH and CT in the maintenance of calcium homeostasis.

Osteoporosis, a bone condition that occurs during and after the menopause, probably involves some disturbances in either one or both of these feedback mechanisms. In normally menstruating women, estrogen does not seem to affect vitamin D production or the circulating levels of PTH and CT (Baran, et al., 1980). The three- to fourfold increase in estrogen which occurs at mid-cycle causes no significant changes in the substances that regulate bone metabolism. See Table 4-2.

During the climacteric, on the other hand, two things are happening, both of which affect the bones. Estrogen production decreases and age-related changes increase. Although the changes due to aging affect bone formation in men as well as in women, they are more serious in women. Why? The answer is probably related to estrogen production, and to the effects of estrogen on the hormones which control bone metabolism.

One of these hormones, calcitonin, decreases with age (Shamonki, et al., 1980). This decrease in calcitonin (CT) affects the incorporation of calcium into bone by the negative feedback mech-

TABLE 4–2 Effect of the Menstrual Cycle on Calcium-Regulating Hormones in Normal Young Women

	*Estrogen (pg/ml)**	*Total Calcium (mg/dl)**	*PTH (μl eq/ml)**	*CT (pg/ml)**	*Vitamin D (pg/ml)**
Day 3	63 ± 9	9.4 ± 0.1	6 ± 1	30 ± 3	47 ± 3
Day 13	229 ± 65	9.6 ± 0.1	7 ± 1	29 ± 4	51 ± 3
Probability of Difference	0.025	N.S.†	N.S.†	N.S.†	N.S.†

*These are units of weight of the substance measured per unit volume of blood plasma.

†N.S. means not significant. The table shows that in normal young women the mid-cycle estrogen elevation is not accompanied by increases in total calcium, PTH (parathyroid hormone), CT (calcitonin), or Vitamin D.

Source: Adapted from D.T. Baran, M.P. Whyte, M.R. Haussler, L.J. Deftos, E. Slatopolsky, and L.V. Avioli, "Effects of the Menstrual Cycle on Calcium-Regulating Hormones in the Normal Young Woman," *J. Clin. Endocr. Metab.* 50, no. 2 (February 1980): 378. © 1980, The Endocrine Society, adapted by permission of the publisher and the author.

anisms illustrated in Figure 4-14(B). Estrogen makes women's bone tissue less sensitive to PTH (Heaney, 1979). Therefore, in the absence of estrogen, bone is more sensitive to PTH and bone resorption occurs faster than bone formation [Fig. 4-14(A)]. The net effect of menopausal changes in CT and PTH is illustrated in Figure 4-15. Increased sensitivity of the bone to PTH and lowered CT production decrease the incorporation of minerals into bone. Menopausal osteoporosis is the result of these disturbances in the negative feedback systems that control bone metabolism.

During the menopausal years, declining estrogen production

FIGURE 4-15 Premenopausal and menopausal bone.

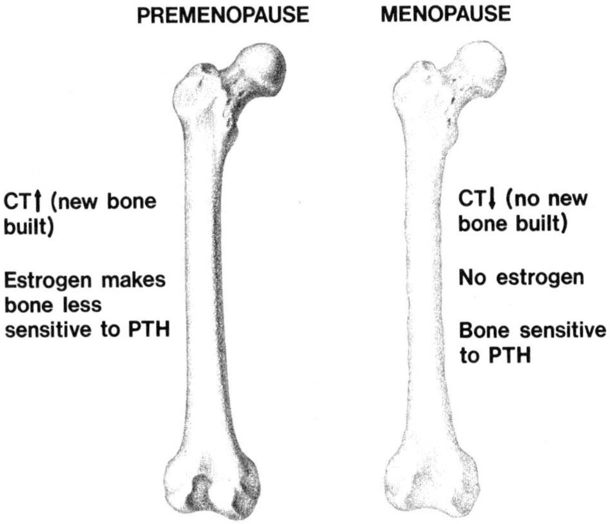

PREMENOPAUSE

CT↑ (new bone built)

Estrogen makes bone less sensitive to PTH

MENOPAUSE

CT↓ (no new bone built)

No estrogen

Bone sensitive to PTH

brings about physiological changes that may be inconvenient, uncomfortable, and possibly even incapacitating. As we have seen, these changes include hot flashes, atrophy of the vaginal epithelium, and osteoporosis. Dealing with the menopause requires that we cope with these changes and integrate them into the pattern of our daily lives.

Dealing with the Menopause

The menopause, like the menarche, is a physiological fact of life. Each of us will have to face this fact in a way that is consistent with our thoughts and feelings about ourselves. These thoughts and feelings may be negative. We may believe that the menopause makes us less of a woman, less feminine, less attractive, less everything. These feelings about the menopause may be based on false expectations—from our society and its concept of a woman's role in it, and from ourselves and what we judge our own role to be.

Hormones, by their presence or absence, may affect our feelings and behavior (see Chapter 7). It is not clear how this happens during the menopause and why some women experience depression, irritability, and anxiety very intensely while others experience them only slightly or not at all. Each of us will have to find her own way, psychologically, of dealing with the menopause.

We live in a society where pill-taking is the easy answer to every complicated problem. Perhaps if we just take pills the menopause will either go away or it will not bother us so much. Do we have a pill for this purpose?

Pills prescribed for the relief of menopausal symptoms contain conjugated estrogens. This estrogen replacement therapy (ERT) has been used for many years to combat the effects of menopausal hot flashes, vaginal atrophy, and osteoporosis. The idea that ERT will make a woman "forever feminine" or eternally young may be extraordinarily appealing. It is also extraordinarily incorrect. If we expect ERT to make us feel psychologically "up" all the time, we are in for a big disappointment. There is no evidence to justify the use of estrogen in the treatment of primary psychological problems. "Nonspecific or psychiatric symptoms should be treated with specific, indicated, nonhormonal medication, if at all. No drug will prevent the natural course of aging; estrogen should not be given in a futile attempt to do so" (Quigley and Hammond, 1979).

Estrogen replacement therapy does improve some physiological effects of the menopause. Estrogens are effective in overcoming the atrophy of the vaginal epithelium with associated symptoms such as dryness, trauma, itching, and pain during intercourse. Women who want to maintain an active sex life may need ERT to achieve this purpose. Creams containing estrogens may be used. They are absorbed rapidly into the bloodstream. No one knows the biological effect of this absorption. For vaginal atrophic changes, ERT can be administered at low levels and therapy may continue for many years. The changes that may occur in the endometrium are similar to those observed in adenocarcinoma of the endometrium (Aycock and Jollie, 1979). A woman who is taking even low doses of estrogens to correct the vaginal atrophy of menopause should have endometrial biopsies done on a yearly basis (N.I.H., 1979). She should take the lowest dose of ERT needed to correct this symptom, and she should know that the risk of endometrial cancer increases from 1 per 1,000 to several times that frequency in women who use ERT.

Although the causes of menopausal hot flashes are only beginning to be understood, the most specific and effective treatment for this problem is ERT. Hot flashes are most severe during the time of the most severe drop in estrogen function; then they begin to decline. The dosage of ERT should be controlled according to the severity of the symptoms.

Estrogen replacement therapy can retard, but cannot replace, bone loss. Because no more than 25 percent of women are at risk for developing *severe* osteoporosis, ERT is not routinely prescribed for all postmenopausal women. In younger women with low estrogen levels, ERT decreases bone loss and fractures (Hammond et al., 1979).

Except for the use of estrogens, most treatments for osteoporosis are still in the investigative stages. In the meantime, some aspects of our lives can be controlled (see Chapter 11) in order to prevent premature thinning of bone tissue. Because inactivity can hasten calcium loss, a regular program of physical activity is recommended. Increased dietary calcium may also help to decrease bone loss, unless individuals have high levels of urinary calcium and/or kidney stones. Unfortunately, most postmenopausal women in this country today consume only one-third as much calcium as they would need to retard bone loss (Marx, 1980).[1]

ERT can be helpful in overcoming some symptoms of the menopause. It is not a major cure-all and it is not without adverse effects. Manufacturers of conjugated estrogens obviously have a large financial interest in the marketing of their product; nevertheless, they enclose Patient Information brochures stating adverse effects for distribution with their products. With this information and the advice of a physician who is familiar with our individual medical histories, each of us needs to decide whether, in her own case, the benefits of ERT outweigh the risks, known and unknown. Current research into ERT effects will be discussed in more detail in Chapter 7.

[1]Without ERT, dietary calcium equivalent to one and one-third quarts of milk, whole, skim or powdered per day is recommended.

REFERENCES

THE FEMALE REPRODUCTIVE CYCLE AND ITS CONTROL MECHANISMS

Chiazze, Leonard, Jr.; Franklin T. Brayer; John J. Macisco, Jr.; Margaret P. Parker; and Benedict J. Duffy. "The Length and Variability of the Human Menstrual Cycle, *JAMA* 203, no. 6 (February 5, 1968): 377–80.

Clark, J. H., and M. K. Zarrow. "Influence of Copulation on the Time of Ovulation in Women," *Am. J. Obstet. Gynecol.* 109, no. 7 (April 1, 1971): 1083–85.

Espey, Lawrence L. "Ovulation as an Inflammatory Reaction—A Hypothesis," *Biology of Reproduction* 22, no. 1 (February 1980): 73–106.

Jöchle, W. "Current Research in Coitus-Induced Ovulation: A Review," *J. Reprod. Fertil.,* Supp. 22 (1974): 165–202.

Matsumoto, Seiichi; Masao Igarashi; and Yu Nagaoka. "Environmental Ovulatory Cycles," *Int. J. Fertil.* 13, no. 1 (January-March 1968): 15–23.

REPRODUCTIVE MECHANISMS IN OUR MALE PARTNER

Bremner, William J., and David M. DeKretser. "The Prospect for New, Reversible Male Contraceptives," *New Eng. J. Med.* 295, no. 20 (November 11, 1976): 1111–15.

Consumer Reports, "Condoms," October 1979, pp. 583–89.

Diller, Lawrence, and Wylie Hembree. "Male Contraception and Family Planning: A Social and Historical Review," *Fertil. Steril.* 28, no. 12 (December 1977): 1271–79.

Ganong, William F. *Review of Medical Physiology.* Los Altos, Calif.: Lange Medical Publications, 1975.

Lindholmer, C. "The Importance of Seminal Plasma for Human Sperm Motility," *Biology of Reproduction* 10 (1974): 533–43.

Masters, William H., and Virginia E. Johnson. *Human Sexual Response.* Boston: Little, Brown, 1965.

Robinson, Derek, and John Rock. "Intrascrotal Hyperthermia Induced by Scrotal Insulation: Effect on Spermatogenesis," *Obstet. Gynecol.* 29, no. 2 (February 1967): 217–23.

World Health Organization. "Reproductive Function in the Male," Technical Report Series no. 520. Geneva: World Health Organization, 1973.

MECHANISMS OF FEMALE SEXUAL RESPONSE

American Medical Association. *Human Sexuality.* Chicago: The American Medical Association, 1972.

Masters, William H, and Virginia E. Johnson. *Human Sexual Response.* Boston: Little, Brown, 1965.

THE MENOPAUSE

Aycock, Nancy R., and William P. Jollie. "Ultrastructural Effects of Estrogen Replacement on Postmenopausal Endometrium," *Am. J. Obstet. Gynecol.* 133, no. 4 (October 15, 1979): 461–66.

Baran, Daniel T.; Michael P. Whyte; Mark R. Haussler; Leonard J. Deftos; Eduardo Slatopolsky; and Louis V. Avioli. "Effects of the Menstrual Cycle on Calcium-Regulating Hormones in the Normal Young Woman," *J. Clin. Endocr. Metab.* 50, no. 2 (February 1980): 377–79.

Chakravarti, S. W.; P. Collins; M. H. Thom; and J. W. Studd. "Relation Between Plasma Hormone Profiles, Symptoms and Response to Oestrogen Treatment in Women Approaching the Menopause," *Br. Med. J.* 1 (1979): 983–85.

Cruess, R. L., and K. C. Hong. "The Effect of Long-Term Estrogen Administration on Bone Metabolism in the Female Rat," *Endocrinology* 107, no. 4 (April 1979): 1188–93.

Frumar, Anthony M.; David R. Meldrum; F. Geola; Issa M. Shamonki; Ivanna V. Tataryn; Leonard J. Deftos; and Howard L. Judd. "Relationship of Fasting Urinary Calcium to Circulating Estrogen and Body Weight in Postmenopausal Women," *J. Clin. Endocr. Metab.* 50, no. 1 (January 1980): 70–75.

Hammond, C. B.; F. R. Jelovsek; and K. L. Lee. "Effect of Long-Term Estrogen Replacement Therapy: 1. Metabolic," *Am. J. Obstet. Gynecol.* 133, no. 5 (March 1, 1979): 525–36.

Heaney, Robert. "Age-Related Changes in Calcium Metabolism in Perimenopausal Women and Their Relation to the Development of Osteoporosis." Conference on the Endocrine Aspects of Aging, Bethesda, Md., October 1979.

Marx, Jean. "Hormones and Their Effects in the Aging Body," *Science* 206, no. 16 (November 16, 1979): 805–6.

———. "Osteoporosis: New Help for Thinning Bones," *Science* 207 (February 8, 1980): 628–30.

Meldrum, David R.; Ivanna R. Tataryn; Anthony M. Frumar; Yohanan Erlik; K. H. Lu; and Howard L. Judd. "Gonadotropins, Estrogens, and Adrenal Steroids during the Menopausal Hot Flash," *J. Clin. Endocr. Metab.* 50, no. 4 (April 1980): 685–89.

Molnar, George W. "Body Temperature during Menopausal Hot Flashes," *J. Appl. Physiol.* 38 (3): 1975: 499–503.

N.I.H. Consensus Development Conference Summary, "Estrogen Use and Postmenopausal Women," *Ann. Int. Med.* 91, no. 6 (December 1979): 921–22.

Perlmutter, Johanna F. "A Gynecological Approach to Menopause," in *The Woman Patient,* ed. Malkah T. Notman and Carol Nadelson. New York: Plenum Press, 1978.

Quigley, Martin M., and Charles B. Hammond. "Estrogen Replacement Therapy—Help or Hazard?" *New Eng. J. Med.* 301, no. 12 (September 20, 1979): 646–48.

Reaves, T. A., and J. M. Hayward. "Hypothalamic and Extrahypothalamic Thermoregulatory Centers," in *Body Temperature Regulation, Drug Effects and Therapeutic Implications,* ed. Peter Lomax and Eduard Schönbaum, pp. 45–47. New York: Marcel Dekker, 1979.

Roth, George S. "Hormone Action and Receptors During Aging." Conference on the Endocrine Aspects of Aging, Bethesda, Md., October 1979.

Schiff, Isaac, and Emery Wilson. "Clinical Aspects of Aging in the Female Reproductive System," in *The Aging Reproductive System,* ed. Edward L. Schneider, pp. 9–29. New York: Raven Press, 1978.

Shamonski, I. M.; A. M. Frumar; I. V. Tataryn; D. R. Meldrum; B. H. Davidson; J. G. Parthemore; H. L. Judd; and L. J. Deftos. "Age-Related Changes of Calcitonin Secretion in Females," *J. Clin. Endocr. Metab.* 50, no. 3 (March 1980): 437–39.

Sherman, Barry M.; Robert B. Wallace; and Alan E. Treloar. "Pathogenic Implications of Menstrual and Hormonal Patterns." Conference on the Endocrine Aspects of Aging, Bethesda, Md., October 1979.

Siiteri, Peneti K. "Interaction of Androgens, Estrogens and the Sex Hormone Binding Globulin." Conference on the Endocrine Aspects of Aging, Bethesda, Md., October 1979.

Timiras, Paola. "Neuroendocrine Strategies to Modify Aging." Conference on the Endocrine Aspects of Aging, Bethesda, Md., October 1979.

Voda, Ann. "Menopausal Hot Flash," Conference on the Endocrine Aspects of Aging, Bethesda, Md., October 1979.

Wallace, R. B.; B. M. Sherman; J. A. Bean; A. E. Treloar; and L. Schlabaugh. "Probability of Menopause with Increasing Duration of Amenorrhea in Middle-Aged Women," *Am. J. Obstet. Gynecol.* 135, no. 8 (December 15, 1979): 1021–24.

5

Pregnancy, Childbirth, and Lactation

During the 35 or so years of our reproductive period, many of us will experience the birth of at least one child. In this chapter, we shall consider the control mechanisms associated with the various stages of pregnancy—including its culmination in labor and delivery—and the mechanisms that govern the period of lactation during which our body supplies milk for our newborn.

THE BEGINNING OF PREGNANCY: FERTILIZATION AND CONCEPTION

We can get pregnant if we participate in unprotected intercourse (no contraceptive method used by either partner) during the time when we are ovulating. At this time the ovum, which has just been released from the ovary, is carried to the oviduct where movements of the cilia propel it upwards. The rate of movement of the cilia depends upon the circulating levels of estrogen and progesterone.

Sperm deposited in the vagina during intercourse move upwards through the uterus to the oviduct. Smooth muscles in the uterus and oviduct contract, and this muscle contraction moves the sperm much more rapidly than they could move by themselves. The route taken by the sperm in their passage through the female reproductive tract is illustrated in Figure 5-1. The time required for this journey to take place has been measured as between 5 and 30 minutes (Settlage, 1973).

Relatively few—probably less than 100—of the vast numbers of sperm deposited in the vagina during intercourse actually reach the site of fertilization. What happens to most of the sperm? Some may pass out of the oviduct and into the body cavity; others may

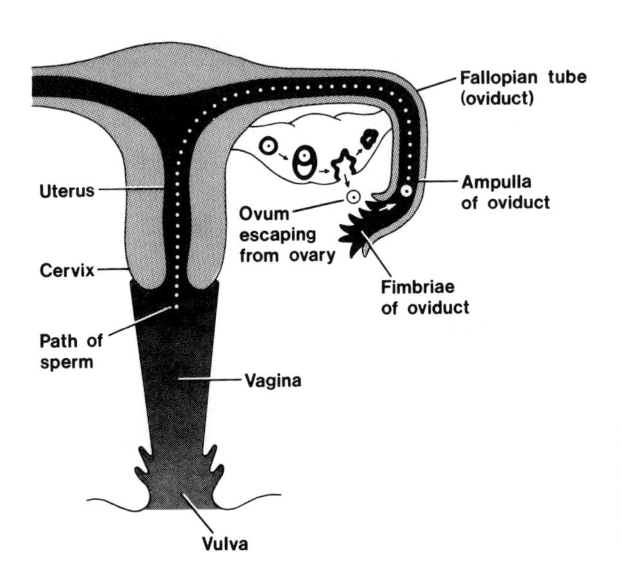

FIGURE 5-1 Movement of sperm cells to the site of fertilization. Adapted from Duffy and Wallace, *Biological and Medical Aspects of Contraception.*

move backward toward the uterus. Within the female reproductive tract, sperm undergo a process that prepares them to fertilize the ovum. This process is called capacitation. After sperm are capacitated, their life span is even more limited. It is possible that most of the sperm are capacitated too early and are unable to fertilize the ovum at the proper time.

"The proper time" is important. The sperm and the ova, the gametes, are delicate and short-lived reproductive cells. The average time of survival for sperm, once it has been deposited in the female reproductive tract, has been estimated by Belonschkin (1949) as follows:

Part of the Tract	*Survival in Hours*
Vagina	2.5
Cervix	48
Uterus	24
Oviduct	48

The upper limit of sperm survival appears to be about 48 hours. Moving sperm have been found in cervical mucus as long as six to eight days after intercourse (Nicholson, 1965). The fact that they are moving, however, does not prove that they still have the ability to fertilize the ovum.

The life span of the unfertilized ovum is even shorter, probably around 24 hours. These time factor limitations affect the possibility of conception even if no contraceptive measures are taken. Since

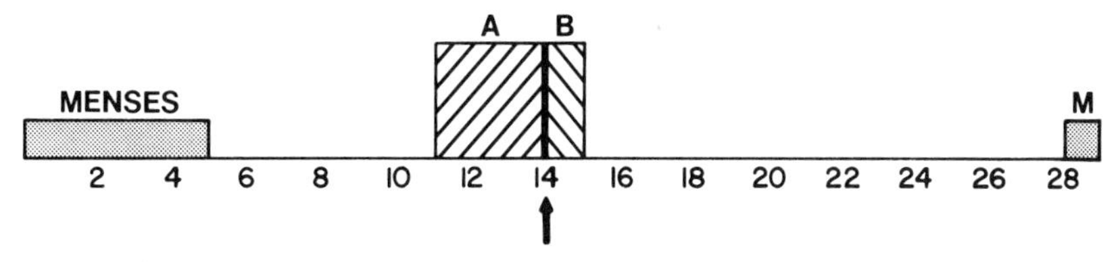

FIGURE 5-2 Length of the fertile period. Section A represents the estimated time of survival for the sperm (two day maximum, plus a one-day safety margin). Section B allows one day for the ovum to remain fertilizable, presuming that ovulation occurred on day 14 of this cycle. Adapted from Duffy and Wallace, *Biological and Medical Aspects of Contraception.*

ovulation usually occurs toward the middle of the cycle, the possibility of pregnancy is greater at that time. Although the day of the ovulatory cycle on which fertilization of an ovum might occur varies—not only from one woman to another but even from one ovulatory cycle to another in the same woman—there are only about five days during any given cycle on which intercourse could result in pregnancy. This so-called "fertile period" is illustrated in Figure 5-2. Intercourse during days 11 to 15 of this cycle could most probably result in pregnancy if no attempts at contraception were made.[1]

The process of fertilization at the cell-to-cell level is similar in all mammals, including mice, rats, cats, dogs, and monkeys, as well as human beings. Fertilization of an ovum by a sperm involves a sequence of events that begins with the capacitation of the sperm and culminates in syngamy.

1. Capacitation of the Sperm. To become capable of fertilizing the ovum, sperm need to spend some time in the secretions of the female genital tract. For a long time scientists believed that this would prevent the in vitro (test tube) fertilization of mammalian ova. In recent years, however, a substance that could substitute for female tract secretions has been identified. The laboratory fertilization of eggs from small mammals is now a routine process because capacitation can and does occur within artificial media.

2. Penetration of the Sperm. In order to penetrate the ovum, the head of the sperm secretes an enzyme that dissolves the substance surrounding the follicle cells. The movements of the sperm are random, and many sperm will usually be found near the ovum. It is in the cell-to-cell contact between ovum and sperm that the sperm's own capacity for movement is important.

[1]Benedict J. Duffy and M. Jean Wallace, *Biological and Medical Aspects of Contraception* (Notre Dame: University of Notre Dame Press, 1969). Reprinted with permission.

3. Attachment of the Sperm. The sperm then attaches itself to the outer membrane of the ovum. This may occur through an antigen-antibody reaction.

4. Fusion of the Sperm. Next, the sperm fuses with the inner membrane of the ovum. After this, the sperm moves into the cytoplasm of the ovum, where the sperm tail disintegrates.

5. Pronuclear Formation. The sperm head, which is really one large nucleus, swells and increases in size. The number of chromosomes within the sperm was already reduced to 23 during sperm formation in the testes. These chromosomes will be the paternal contribution to the zygote. The maternal contribution, besides the chromosomes, includes the egg cytoplasm, which will furnish developmental material for subsequent processes.

6. Syngamy. Syngamy (from Greek *syn* 'together', and *gamos* 'marriage') is the essential part of fertilization. The 23 chromosomes in the sperm unite with the 23 chromosomes in the ovum. The zygote, formed by the union of sperm and ovum, contains 46 chromosomes. This inherited chromosomal information will enable it to follow the orderly sequence of events known as development.

The stages in the process of fertilization are diagrammed in Figure 5-3.

Fertilization activates the ovum so that it begins to divide. As cell division proceeds, the zygote moves down the oviduct to the body of the uterus. By the third day after fertilization the embryo is in the uterus, where it is nourished by the glands in the endometrium.

Six days after fertilization, the embryo consists of two parts, an inner part which will develop into the embryo then into a fetus, and an outer layer of cells which contacts the endometrium. Enzymes secreted by the outer cell layer penetrate into the endometrium, where the developing embryo absorbs food until it develops contact with the maternal circulatory system.

The implantation of the embryo in the endometrium, illustrated in Figure 5-4, depends upon the coordination of several control mechanisms:

1. *in the male, production of adequate numbers of normal spermatozoa, erection, and ejaculation*
2. *in the female, ovulation and the preparation of the endometrium*
3. *timing of intercourse*

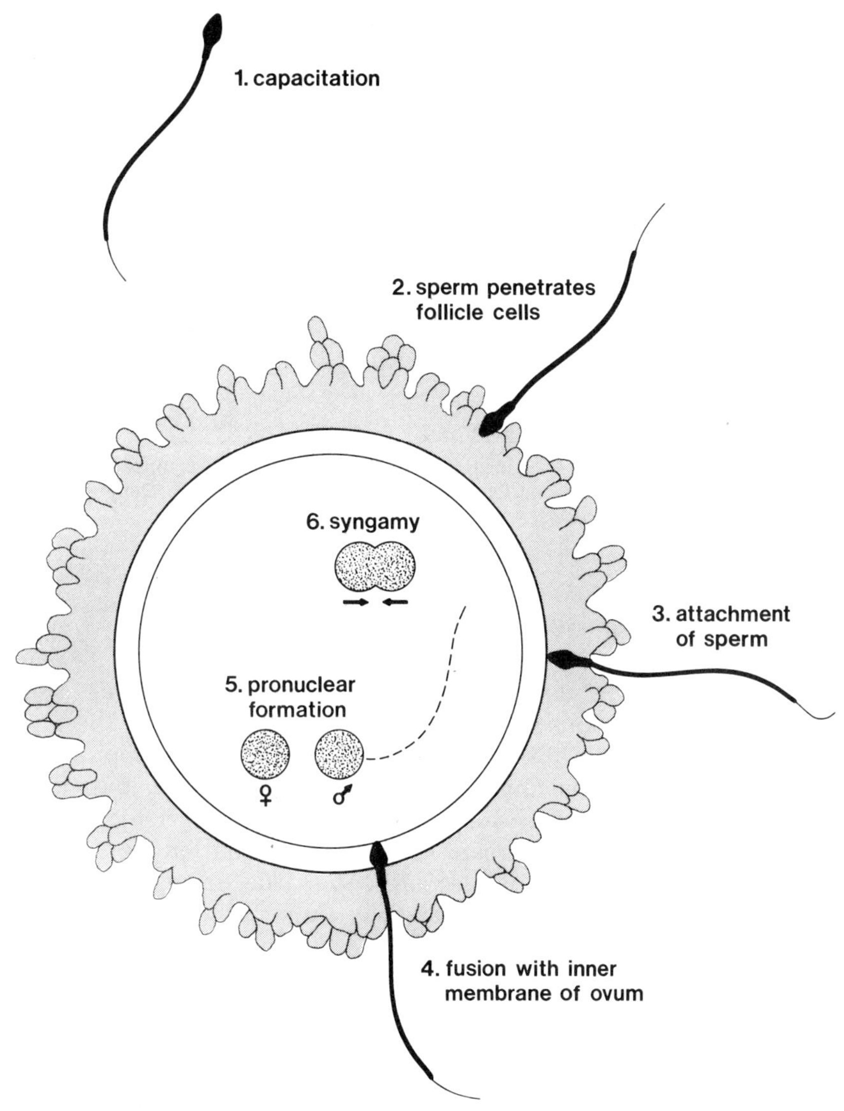

FIGURE 5-3 Summary of events at the cellular level during fertilization.

4. *hormonal control of gamete movement*
   a. *sperm upwards*
   b. *ovum to uterus at the time the endometrium is ready to receive it.*

All of these control mechanisms operate well in most individuals. In others, however, unwanted infertility (see Chapter 6) is a result of some failure in the control mechanisms.

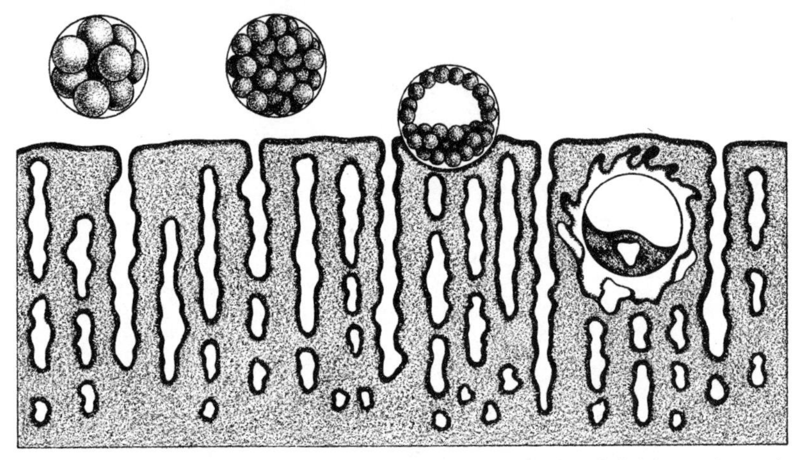

FIGURE 5-4 Implantation. The zygote undergoes repeated cell division as it travels down the oviduct into the uterus. Adapted from Duffy and Wallace, *Biological and Medical Aspects of Contraception.*

CONTROL MECHANISMS FOR PREGNANCY

How does the ovary know that the uterus is pregnant? How do *we* know that we are pregnant? Both of these questions have the same answer: human chorionic gonadotropin (HCG). When the zygote is implanted in the endometrium, the tissues involved begin to secrete this hormone. Human chorionic gonadotropin travels in the bloodstream and inhibits ovulation for the duration of the pregnancy. It is also present in the urine, as well as in the blood. Its presence in either body fluid is the basis for commonly used pregnancy tests.

Pregnancy Tests

A missed menstrual period, feelings of nausea, weight gain, and increases in body temperature may be early subjective signs of pregnancy. A sexually active woman in her reproductive years may experience these symptoms. Ordinarily, these signs need to be confirmed by a pregnancy test. The validity of the pregnancy test depends on the timing of the test. We may have only a general idea about when our missed menstrual period should have occurred. Our knowledge of the variability in our own cycles may be even less precise. Apprehension over an undesired pregnancy or an overeagerness to become pregnant could result in a pregnancy test that is falsely negative because it was done too soon.

Levels of HCG in the blood and urine begin to rise soon after

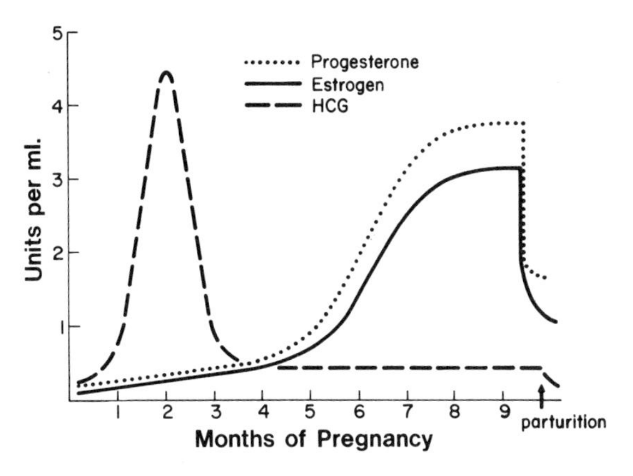

FIGURE 5-5 Hormone levels during pregnancy.

the implantation of the embryo. As illustrated in Figure 5-5, they reach a peak at between the eighth and tenth weeks of pregnancy, then fall to lower levels as pregnancy proceeds. Pregnancy tests for the presence of HCG in the blood or urine are based on reactions that bind the hormone to receptors or to antibodies. The sensitivity of the measuring method (assay) determines how soon the HCG can be measured. For example, a measuring method may detect HCG when it is present at moderately elevated levels. If this amount is present in most women about two weeks after a missed menstrual period, then the method would be an accurate test for pregnancy four weeks after fertilization or six weeks after the last menstrual period (LMP).

The LMP is an important date. It refers to the first day of the last menstrual period. Usually it is easier to remember this date than the date on which intercourse resulted in the fertilization of the ovum. The time of gestation is calculated from the LMP. Figure 5-6 shows two different ways of calculating the total time of pregnancy: gestation time and development time.

FIGURE 5-6 Duration of pregnancy: Gestation time and development time.

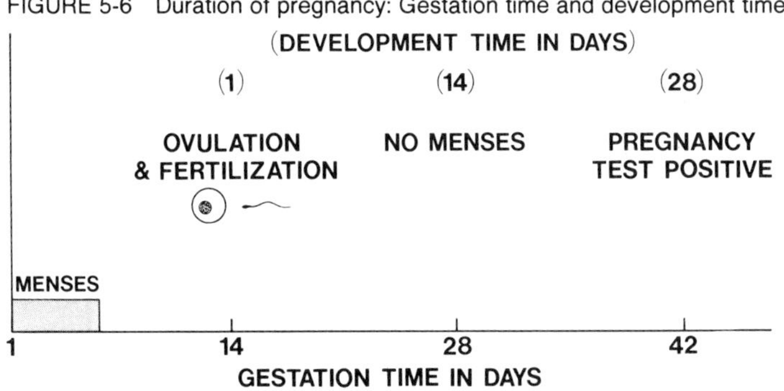

If we think we may be pregnant because we have some external signs of pregnancy, and we weren't using any form of contraception, we have several options with regard to pregnancy testing.

"Do-it-yourself" pregnancy tests are a very appealing idea for many of us. Finding out for ourselves, rather than waiting to be told, seems to put us more in control of the situation. The pregnancy test kits available for home use sell over the counter in drug stores for $7 to $10. The directions, which suggest that the test can be used as early as nine days after the missed menstrual period, must be followed *exactly*. The manufacturers of the tests claim varying degrees of accuracy, some as high as 97 percent on a positive test. But what if the test is negative? The manufacturers' claim of 80 percent accuracy at this end of the spectrum may be too high if a woman uses it too early, or if, in her case, the levels of HCG are not yet high enough to be measured by the sensitivity of the test. Table 5-1 shows the results from several different laboratory slide tests for pregnancy. Days 29 to 42 on this table would represent 1 to 14 days after the missed period. Note that, at this time, only one-third of the tests are positive, and one is falsely positive. The results are much more reliable when the tests are performed later, more than 15 days after the missed period.

A woman who thinks she is *not* pregnant when in fact she *is* faces danger from several sources. She could have a pregnancy in which the embryo is developing in her oviduct or abdominal cavity. Because the embryo is not developing within the uterus, HCG will not be detected by the pregnancy test. This type of pregnancy, called an ectopic pregnancy, is always dangerous. In a physician-directed pregnancy test, the possibility of ectopic pregnancy would have been eliminated by a pelvic examination. A woman who thinks she is not pregnant may also continue to use drugs, alcohol, and medications that could harm the embryo.

In spite of their drawbacks, self-administered pregnancy tests may be useful to some of us in certain circumstances if they are used correctly. There are other more reliable and less expensive alternatives. If we have a private physician, we do not need to wait

TABLE 5–1 Results of Laboratory Slide Tests for Pregnancy

Day of LMP	*% positive*	*% negative*	*% pregnant*
0–14	0	100	0
15–28	0	100	0
29–42	33	67	21
43–56	93	7	93
57–70	92	8	92

Source: Thomas F. Sullivan, William F. Borg, Jr., and Gerald Stiles, "Evaluation of a New Rapid Slide Test for Pregnancy," *Am. J. Obstet. Gynecol.* 133, no. 4 (February 15, 1979): 414, by permission of C. V. Mosby Company and the authors.

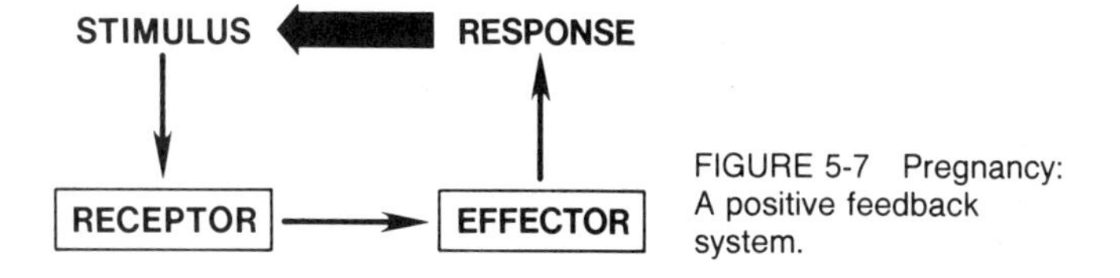

FIGURE 5-7 Pregnancy: A positive feedback system.

to see her before having the pregnancy test done. When we call her office, we will be told where to take the urine sample for an analysis. Usually, we will be instructed to take the sample of urine from the first morning voiding to a laboratory. The laboratory will have the result by the end of the day, and we may obtain the result by calling, not the laboratory, but the doctor's office. Many state, county, and city health departments offer inexpensive and confidential pregnancy testing, usually as part of their family planning programs. The Planned Parenthood Federation of America, Inc., has over 700 centers in the United States. Typical fees are $5 to $10 and may be less for those who cannot afford this amount. Results are available promptly.

If we think we may be pregnant and have a good reason for wanting results before the usual kinds of pregnancy tests will be positive, we may inquire about having a pregnancy test done by one of the newer methods of hormone measurements, the radioimmunoassays and radioreceptor assays.[2] The radioisotope assays can detect levels of HCG present *before* the missed menstrual period, at least two weeks before regular pregnancy tests are accurate. These tests are more complicated to perform. They require specially trained personnel and expensive equipment which are usually found at medical centers. At the present time radioimmunoassays and radioreceptor assays are not used for the routine diagnoses of uncomplicated pregnancies.

Pregnancy as a Positive Feedback System

A positive pregnancy test and a positive pregnancy put our bodies into a state that is characterized by positive feedback systems. The ovarian hormones, estrogen and progesterone, are produced until the end of pregnancy (Fig. 5-5). The growth of the embryo also continues without any negative feedback regulation. These two interdependent events, hormone production and embryo-fetal development, operate as positive feedback systems, in which the response augments, rather than shuts off, the stimulus. See Figure 5-7. As the embryo grows, the corpus luteum produces more and more estrogen and progesterone. After the eighth week of gestation, the

[2]Radio- implies the use of radioactive isotopes in the measuring system.

embryo is referred to as a fetus. By this time, the genes that direct its development have shown that they are human genes on human chromosomes. As the pregnancy proceeds, obvious changes take place in our bodies. The changes that take place in the embryo/fetus are even more profound. Table 5-2 indicates how changes in the growing fetus are correlated to changes in the mother's body. Figure 5-8 shows the relative size of the fetus at various stages in its development.

At the same time the fetus is growing, another structure is growing within the uterus, the placenta. This remarkable structure develops from the place at which the embryo contacts the endometrium. At first the embryo is nourished by materials from the endometrium itself. By the end of the second week of pregnancy,

TABLE 5–2 Development Chart of Pregnancy

Week	*Embryo/Fetus*	*Mother*
1–2	Microscopic—small group of cells.	Misses menstrual period.
3	Limbs begin to form. Embryo 1/10 inch in length.	
4	Eyes, ears, nose develop. Embryo 1/5 inch in length.	Breasts enlarge. No ovulation. Pregnancy test is positive.
8	Limbs appear as distinct. *Fetus* 1 inch in length.	
12	Sex differentiation begins. Fingers and toes appear. Length, 3 inches.	Breasts change; nipples have more pigment. Uterus is above pelvic bones.
16	Sex differences clear. Length, 5 inches; weight 1/4 pound.	Abdomen protrudes. Some fetal movement may be detected. Placenta well-formed.
20	Hair appears. Length, 8 inches; weight 3/4 pound.	Uterus expanded to navel.
24	Length, 12 inches; weight, 1 1/2 pounds.	Strong fetal movements may be felt.
28	Length, 15 inches; weight, 3 pounds.	Uterus 4 inches above navel.
32	Length, 16 inches; weight, 4 pounds.	Uterus approaches breast bone.
36	Length, 18 inches; weight, 5 pounds.	Uterus reaches breast bone. Irregular uterine contractions. Cervix thins. Head of fetus descends to true pelvis.
40	Term—ready to be born. Length, 20 inches; weight, 7 to 8 pounds.	

Source: Jack M. Futoran and May Annexton, *Your Body: A Reference Book for Women* (New York: Ballantine Books, 1976), pp. 85–87, adapted by permission of Jack M. Futoran.

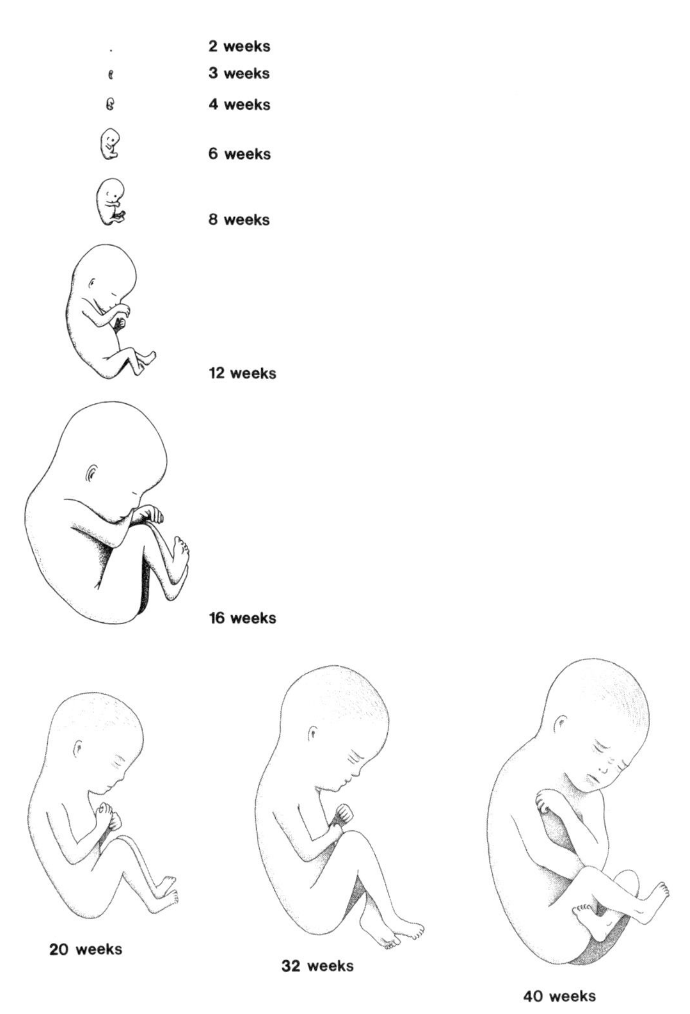

FIGURE 5-8 Embryonic/fetal development during pregnancy. The figures at 2 through 16 weeks show the embryo/fetus at slightly more than half its actual size.

extensions from the outer part of the embryo have made contact with the mother's circulatory system. When the embryonic heart begins to pulsate (about 26 days after fertilization), the circulation of the embryo's blood begins. Through extension into the endometrium, the embryonic/fetal blood passes in close contact to the maternal blood. The two circulatory systems are not in *direct* contact, but they exchange products throughout the pregnancy.

As the fetus grows, two blood vessels, the umbilical artery and the umbilical vein, increase in length. They are contained in the umbilical cord, which connects the fetus to the placenta. At term, the fully developed placenta, with the umbilical cord, weighs about one-sixth as much as the infant. Its size is indicative of its important functions. During intrauterine life the placenta serves as a lung, kidney, and digestive system for the fetus. How does it fulfill these functions?

The fetus is encased within a sac, the amnion. The fetus does not breathe air or digest food; it is completely dependent upon the food and oxygen that are brought to it by the mother's circulatory system. Carbon dioxide and other waste products of fetal metabolism are returned, via the umbilical vein, to the maternal circulation to be excreted by the mother's lungs and kidneys, respectively.

The placenta also functions as an endocrine organ. It secretes estrogen and progesterone in increasing amounts as the pregnancy progresses. Both hormones are required to maintain the pregnancy. The functions of estrogen during pregnancy include the enlargement of the breasts and growth of glandular tissue, the enlargement of the uterus and vulva, and the relaxation of ligaments in the pelvis. Under the influence of estrogen, the anterior junction of the pelvis, the symphysis pubis, becomes more elastic in preparation for the passage of the fetus through the birth canal. Progesterone is important because it assists in preparing the breasts for lactation. Throughout the pregnancy, progesterone decreases the contractability of the uterine muscle and prevents spontaneous abortion. The placenta thus takes over the functions of the corpus luteum, including the production of HCG, so that the corpus luteum is no longer needed after about the third month of pregnancy. The placenta becomes the site of the principal control mechanisms that operate during pregnancy: hormone production and fetal growth.

The development indicated in Table 5-2 and Figure 5-8 define a normal pregnancy. Some variations will be within the normal range of pregnancy events. Abnormal pregnancy events, such as ectopic pregnancy, require surgical interruption of the pregnancy. Some uterine pregnancies may be terminated before the birth of the infant; this is referred to as abortion (see Chapter 6).

Pregnancy and Maternal Health

We can, through certain choices that we ourselves control, increase our chances of having a normal pregnancy and giving birth to a healthy baby. Although some aspects of our general health lie outside our control, many equally important factors do not. For those of us with genetically determined hormone imbalances or enzyme deficiencies that interfere with our metabolic processes, getting pregnant in the first place was a difficult process. But, once pregnant, we have some choices that will improve the general health of our babies.

The dependence of the fetus on the mother, via the placenta, is the basis for urging the pregnant woman to maintain herself in

TABLE 5-3 Some Outcomes of Maternal Smoking during Pregnancy

Outcome Affected	*Effect of Smoking*	*Reference*
Survival of fetus	Perinatal mortality increases	Meyer and Tonascia, 1977
Fetal breathing movements	Decreased	Manning et al., 1975
Spontaneous abortion	Increased	Kline et al., 1977
Behavior of infant	Hearing impaired	Saxton, 1978
Circulatory system	Blood vessel changes	Asmussen, 1978

good health and to select a nutritious diet. The food she eats and digests passes from her blood to that of the fetus through the placenta.

The excessive intake of drugs such as nicotine and alcohol may retard fetal development of the central nervous system, bones, and other tissues. The effects of excessive alcohol consumption on fetal growth during pregnancy (Fetal Alcohol Syndrome) are considered in Chapter 10. The use of drugs or of any medications during pregnancy has been very strictly curtailed by physicians, particularly since it was found several years ago that thalidomide, a sedative given to pregnant women, produced gross developmental abnormalities in the fetus.

Smoking cigarettes is a freely chosen form of behavior, up to a point. When the time comes that we are no longer free to choose whether or not we will smoke—that is, we *have* to smoke—we have an additional problem. Our problem, if we are pregnant, becomes a problem for the fetus. Well-documented studies[3] have shown that maternal smoking is related to fetal/infant problems, including spontaneous abortion, low birth weight, respiratory problems, obstruction of umbilical blood vessels, and some behavioral problems. Some of these effects of smoking are summarized in Table 5-3. The growing body of evidence linking maternal smoking to increased perinatal mortality is a strong argument. This evidence should be presented to women of childbearing years, so that they know the potential risk of smoking on the outcome of pregnancy (Fielding, 1978).

Many couples want to know whether intercourse during pregnancy, particularly during the later phases, will be harmful to the health of the mother and child. At this time we have no data upon which to base an answer to that question. Intercourse and orgasm during the last three months of pregnancy might be harmful if the woman has a poor reproductive history (Herbst, 1979). More statistics are needed, but sexual expression through intercourse during the last trimester of pregnancy may not be feasible in individual cases. Couples might investigate other ways of showing affection and giving each other pleasure during that time.

[3]The interested reader is referred to *Bibliography on the Effects of Smoking in Pregnancy,* which can be obtained from Reproduction Research Information Services, Ltd., 141 Newmarket Road, Cambridge, CB5, 8HA, England.

PARTURITION (LABOR)

The date on which we are told to expect our baby usually depends, as we have seen, on our ability to remember the first day of the LMP. If we remember that date correctly, then the "due date" will be approximately correct. However, in individual cases, control mechanisms that initiate parturition (labor, childbirth) may be influenced by genetic differences, so that our babies might come before or after the usual time of gestation.

In spite of intensive research and some speculation, the stimulus that initiates parturition remains obscure. Once parturition begins, however, it develops, under normal circumstances, as a cascade phenomenon. In this case, several positive feedback systems operate together to bring about the expulsion of the fetus, the fetal membranes, and the placenta.

Some time in late pregnancy the myometrium, the muscular lining of the uterus, becomes more sensitive to oxytocin. This hormone, released from the posterior part of the pituitary gland, stimulates contraction of the uterus. Uterine contractions have been prevented, as we have seen, by progesterone. It has been suggested that parturition begins when a stimulus of unknown origin blocks progesterone production.

The mechanical, environmental, and hormonal changes that occur at the time of childbirth may either be the stimulus, or augment its effects. Some of these changes are listed below and summarized in Figure 5-9.

1. *Mechanical: Cervix is dilated by the head of the fetus. Nerve impulses travel to the hypothalamus, and stimulate release of oxytocin from the pituitary.*
2. *Environmental: Stress on behalf of the fetus or mother could initiate the release of oxytocin, via nervous and hormonal control mechanisms.*

FIGURE 5-9 Parturition: A positive feedback system.

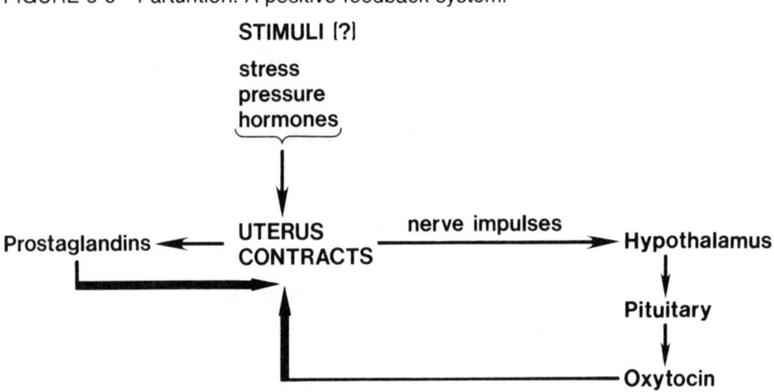

3. *Hormonal:*
   a. *Estrogen. Increases uterine contractility.*
   b. *Progesterone. Decreases uterine contractility.*
   c. *Oxytocin. Increases uterine contractility.*
   d. *Prostaglandin. Increases uterine contractility. (Prostaglandin is not always considered to be a hormone. It is produced by many tissues, including the uterus.)*

The ejection of the fetus from the uterus occurs by means of cervical dilation and uterine contraction. Although the initiating stimulus, as we have seen, is unknown, uterine contractions operate as a positive feedback control mechanism until the fetus and placenta come out of the uterus. The undefined stimulus initiates the three stages of labor.

First Stage of Labor. The first stage of labor lasts from the beginning of regular labor contractions until the cervix is fully dilated and flush with the vagina to form the birth canal. In primagravidas (women pregnant for the first time), this stage usually lasts 10 to 12 hours. In multiparas (women who have given birth at least once before) it lasts 6 to 8 hours.

Second Stage of Labor. The second stage of labor is the period from complete dilation of the cervix until the delivery of the infant. During this stage the strong, involuntary contractions of the uterine muscles can be assisted by pressure from the abdominal muscles. These muscles are under voluntary control. Women who participate in natural childbirth classes learn how to control these voluntary muscles in order to assist with this stage of labor. It is in this stage of labor that an episiotomy may be performed. This incision in the perineum facilitates the movement of the fetus through the vulva. Episiotomies hurt! In primagravidas the second stage lasts about 1½ hours. In multiparas it lasts only 15 to 30 minutes.

Third Stage of Labor. The third and final stage of labor extends from the delivery of the fetus until the delivery of the placenta ("afterbirth") and its membranes. In most women this stage lasts 15 to 30 minutes.

Methods of Delivery

We cannot tell exactly why and when labor may begin to terminate a pregnancy in individual cases. We may, however, have some choice about some of the circumstances of the delivery.

NATURAL CHILDBIRTH Methods of natural childbirth stress the voluntary participation of mother and father in the parturition process.

Couples attend classes to learn about this process. Breathing exercises and the contraction of voluntary muscles—techniques for which are practiced in class—can help in the second stage of labor during an uncomplicated delivery. In a successful natural childbirth, the need for anesthesia is minimal.

Occasionally, however, the woman who is prepared for natural childbirth will find she needs some pain medication during labor. Labor is a painful process. In individual cases it may be more painful because it lasts longer. We should discuss anesthesia during labor with our physician before the event, so that we know what to expect during the immediate postpartum period. Different forms of anesthesia may affect us and the baby in different ways.

CESAREAN SECTION Some of us may need a Cesarean section (C-section) in order to ensure the birth of a normal baby. The reasons for performing this operation include maternal factors such as previous Cesarean section, older primipara, toxemia of pregnancy,[4] infection with venereal disease, and diabetes. Other factors include disproportion (the head of the fetus is too large for the birth canal), malpresentation (the fetus is approaching the birth canal feet first rather than head first), uterine disfunction (the muscles of the uterus don't contract as they should), and a broken placenta.

During the Cesarean section operation, incisions are made through both sets of muscles, the abdominal skeletal muscles and the uterine muscles. The cut muscles, as well as the other cut tissues of the uterus and abdomen, heal, but they cannot contract as forcefully as the unsevered muscle. That is why women who have had one Cesarean section generally have to have all their subsequent children this way.

INDUCTION OF LABOR Sometimes the normal stimuli for the onset of labor, whatever they are, do not begin when they should. After waiting for a reasonable time, our obstetrician may induce labor.

Labor can be induced in several ways. The obstetrician can inject oxytocin, the hormone that stimulates uterine contraction. The prostaglandins, which intensify uterine contraction during childbirth, are also used to stimulate the uterus and to dilate the cervix (Lauersen, 1979). Other methods include amniotomy (the rupture of the sac-like structure enclosing the fetus), and stripping of the membranes (separating the amnion from the wall of the uterus without rupturing it).

Toward the end of our pregnancy, when we visit our obstetrician

[4]Decreases in kidney blood flow and filtration occur in about 1 percent of pregnant women. The mechanisms that cause toxemia of pregnancy are unknown. Rapid weight gain, edema, and elevated blood pressure characterize this condition.

for routine checks, we should discuss these possible adjuncts to labor, with all their effects and side effects. He or she will tell us what to expect if any complications in the delivery of the baby make it necessary to use surgical or hormonal methods in our individual case.

LACTATION

The decision about whether or not we are going to breast-feed has to be made before the baby is born. The control mechanisms that regulate lactation will operate after childbirth unless they are overridden by other means. If we decide not to nurse the baby we are given large amounts of hormone, usually estrogen, in the form of a pill or an injection right after delivery. How does the hormone work to transfer the feeding of the infant from the breast to the bottle?

We have seen (Chapter 3) that the development of the breasts during puberty is stimulated by estrogen and progesterone. During pregnancy, both hormones cause increased development of the duct system in the mammary glands. Another hormone, human placental lactogen, which is secreted by the placenta, also stimulates duct development in the mammary gland. The rich duct system of the breast during pregnancy is illustrated in Figure 5-10.

It is important for us to realize that the size of our breasts has nothing to do with our ability to nurse a baby. The glandular tissue, which produces the milk, develops during the pregnancy. Women with small breasts are just as capable of nursing their babies as those with larger deposits of fat on the mammary glands.

After parturition the sudden drop in progesterone cancels the effect of an inhibitory substance, prolactin inhibition factor (PIF), which is produced by the hypothalamus. When the influence of PIF is negated, the pituitary gland can release prolactin, a hormone that stimulates milk secretion in the mammary glands (Fig. 5-11). When women who decide not to nurse are given large doses of estrogen, the production of milk is probably inhibited by a feedback effect

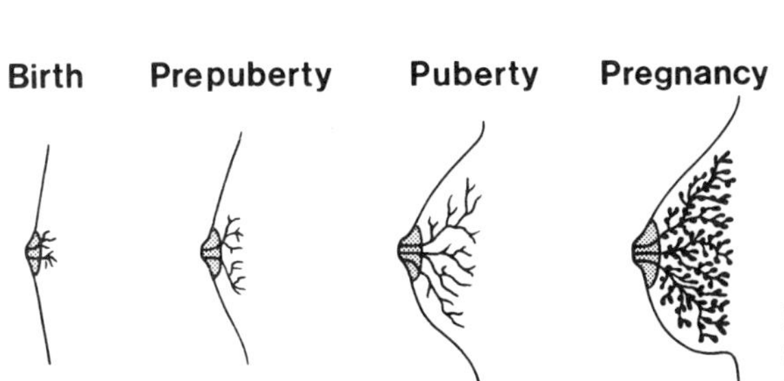

FIGURE 5-10 Development of the mammary glands.

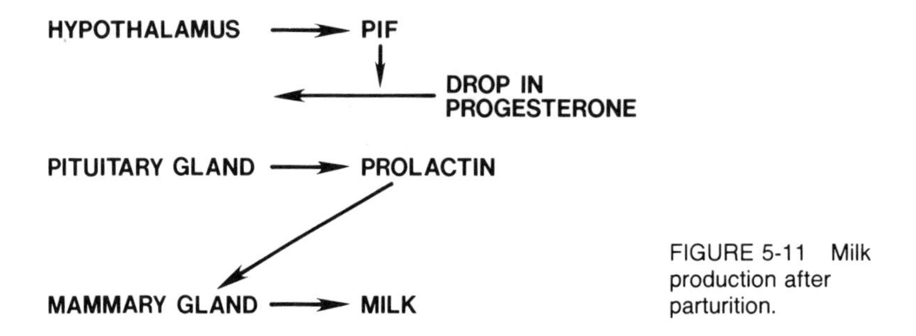

FIGURE 5-11 Milk production after parturition.

of estrogen on prolactin. Milk production requires not only prolactin, but other hormones such as insulin, cortisol, and thyroxine.

We've decided to nurse the baby and the milk is there. How does it get out? "The suckling reflex" is the answer. This reflex, whose mechanism is illustrated in Figure 5-12, is one that depends on both the baby and the mother. The stimulus for this reflex is the baby's suckling on the nipple. Nervous impulses are transmitted to the hypothalamus, where neurosecretory cells deliver a message to the pituitary: "Secrete oxytocin!" This oxytocin is the same hormone that produces smooth muscle contractions in the uterus during childbirth. Its role in the nursing process is to stimulate contraction of the myoepithelial cells, smooth muscle-containing cells that encircle the milk ducts. Milk let-down, or the transfer of the milk from ducts to nipples, is the result of myoepithelial cell contraction. Like the orgasmic reflex, the suckling or nursing reflex can be inhibited by inputs from the brain. Nervousness, fear of failure, or preoccupation with other things can interfere with the reflexive mechanism.

As long as there is a suckling infant, and the mother is ade-

FIGURE 5-12 Milk let-down and milk production during lactation.

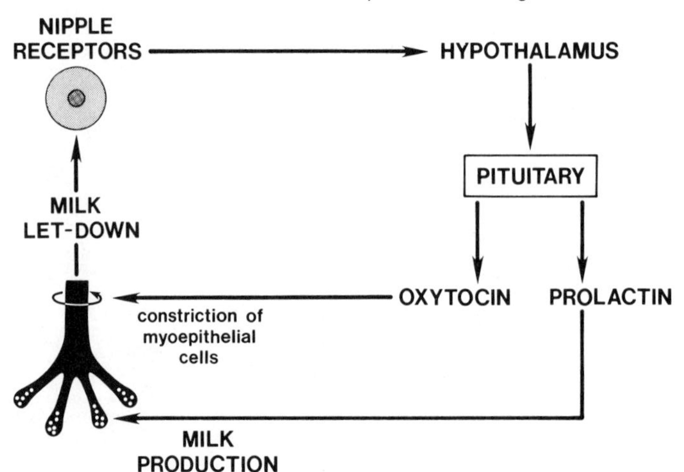

quately nourished, there will be a milk supply. Suckling continues the inhibition of PIF. Prolactin is released from the pituitary gland and milk is formed in the mammary glands as long as the baby continues to nurse. Suckling also inhibits the release of FSH and LH and continues the block of ovulation which was begun during pregnancy. The duration of this lactation-induced sterility varies greatly among individuals. Since ovulation may occur before or after the first postpartum menses, it is not a good idea to rely on lactation for contraception.

Lactation, or breast-feeding the infant, thus depends on the control mechanisms that regulate the development of milk-producing tissues in the breast, the initiation of milk production after delivery, milk let-down, and the maintenance of milk production. Our decision to nurse our baby by allowing these mechanisms to operate is accompanied by benefits. The baby has a good sanitary supply of food which, because it contains the mother's antibodies, confers a certain amount of protection against disease. Human milk is the best food for human babies. Nursing a baby, we may find, is an emotionally satisfying experience. Another benefit to us comes from the oxytocin released during the suckling reflex. It helps contract the uterus to its nonpregnant size so that we are able to get back into our nonmaternity clothes sooner. Nursing is convenient in that the supply of milk is maintained, at the proper temperature, as long as we continue nursing. During pregnancy, a certain amount of fat is stored in our bodies as a prospective nutrient source for nursing. The fat stores are usually used up if a woman nurses, but she may have to diet after the baby is born if she does not nurse.

Nursing our babies may be inconvenient because of social and/or economic pressures. Not all hospitals are set up for the demand feeding schedules that breast-fed babies need. We ourselves will probably need help and reassurance at the start, even though lactation is a natural outcome of pregnancy and parturition. With some help and guidance, it is estimated that 95 percent of women are capable of successful breast-feeding.

REFERENCES

THE BEGINNING OF PREGNANCY: FERTILIZATION AND CONCEPTION

Belonschkin, B. *Zeugung beim Menschen in Lichte der Spermatozoenlehre.* Stockholm: Sjobergs Forlag, 1949.

Gwatkin, Ralph B. L. *Fertilization Mechanisms in Man and Mammals.* New York: Plenum Press, 1977.

Nicholson, R. "Vitality of Spermatozoa in the Endocervical Canal," *Fertil. Steril.* 16 (1965): 758.

Settlage, D. S. F.; M. Motoshima; and D. R. Tredway. "Sperm Transport from the External Cervical Os to the Fallopian Tubes in Women: A Time and Quantitative Study," *Fertil. Steril.* 24, no. 9 (September 1973): 175–79.

Wallace-Haagens, Mary Jean; Benedict J. Duffy, Jr.; and Hugh R. Holtrop. "Recovery of Spermatozoa from Human Vaginal Washings," *Fertil. Steril.* 26, no. 2 (February 1975): 175–79.

CONTROL MECHANISMS FOR PREGNANCY

Asmussen, Inger. "Arterial Changes in Infants of Smoking Mothers," *Postgraduate Medical Journal* 54 (March 1978): 200–204.

Consumer Reports, "Test Yourself for Pregnancy," vol. 43 (November 1978): 644–45.

Fielding, Jonathan E., and Pearl K. Russo. "Smoking in Pregnancy," *New Eng. J. Med.* 298 (February 9, 1978): 337–39.

Frederick, Jean, and Anne B. M. Anderson. "Factors Associated with Spontaneous Pre-Term Birth," *Br. J. Obstet. Gynaecol.* 83 (May 1976): 342–50.

Futoran, Jack M., and May Annexton. *Your Body: A Reference Book for Women.* New York: Ballantine Books, 1976.

Herbst, Arthur. "Coitus and the Fetus," *New Eng. J. Med.* 301, no. 22 (November 29, 1979): 1235–36.

Kline, Jennie; Zena A. Stein; Mervyn Susser; and Dorothy Warburton. "Smoking: A Risk Factor for Spontaneous Abortion," *New Eng. J. Med.* 297 (October 13, 1977): 793–96.

Manning F.; E. Wyn Pugh; and K. Boddy. "Effect of Cigarette Smoking on Fetal Breathing Movements in Normal Pregnancies," *Br. Med. J.* (March 8, 1975): 552–53.

Meyer, M. B., and J. A. Tonascia. "Maternal Smoking, Pregnancy Complications, and Perinatal Mortality," *Am. J. Obstet. Gynecol.* 128, no. 5 (July 1, 1977): 494–502.

National Institute of Neurological Disease and Stroke. *The Women and Their Pregnancies.* DHEW publication no. (N.I.H.) 73-379. U.S. Department of Health, Education and Welfare. Washington, D.C., 1972.

Saxton, David W. "The Behavior of Infants Whose Mothers Smoke in Pregnancy," *Early Human Development* 2, no. 4 (1978): 363–69.

Seligman, Jean. "Home Tests for Pregnancy," *Newsweek,* September 3, 1979, p. 69.

Sullivan, Thomas F.; William F. Borg, Jr.; and Gerald Stiles. "Evaluation of a New Rapid Slide Test for Pregnancy," *Am. J. Obstet. Gynecol.* 133, no. 4 (February 15, 1979): 411–14.

PARTURITION AND LACTATION

Feldman, Silvia. *Choices in Childbirth.* New York: Grosset and Dunlap, 1978.

Findlay, A. L. R. "The Control of Parturition," *Res. Reprod.* 4, no. 5 (September 1975).

———. "Lactation," *Res. Reprod.* 6, no. 6 (November 1974).

Friedman, Emanuel A.; Marlene R. Sachtleben; and Ann K. Wallace. "Infant Outcome Following Labor Induction," *Am. J. Obstet. Gynecol.* 133, no. 6 (March 15, 1979): 718–21.

Lauersen, Niels H.; Steven Seidman; and Kathleen Wilson. "Cervical Priming Prior to First Trimester Suction Abortion with a Single 15-Methyl-Prostaglandin F_{22} Vaginal Suppository," *Am. J. Obstet. Gynecol.* 135, no. 8 (December 15, 1979): 1116–18.

World Health Organization. "Endocrine Regulation of Human Gestation," Technical Report Series no. 471. Geneva: 1971.

6
Contraception, Infertility, and Abortion

During our reproductive years, most of us, if we are sexually active, will employ one or more methods of contraception to avoid unwanted pregnancies, or to delay pregnancy until the time we have chosen for childrearing. While the prevention of pregnancy is a major concern for large numbers of women, others of us face the opposite problem: We want to conceive a child but are unable to do so for reasons that may involve disorders of our own or our partner's reproductive mechanisms. Occasionally, a pregnancy that has begun will be terminated—either because of our own decision, or because of maternal or fetal disorders that prevent our carrying it to completion. The present chapter will discuss the control systems and other issues associated with contraception, infertility, and abortion.

CONTRACEPTION

History

Contraception, the prevention of conception, was a human concern long before the age of modern science. An Egyptian papyrus dated earlier than 500 B.C. describes in detail the contraceptive effect of crocodile dung and honey inserted as a cervical plug prior to intercourse. Biblical quotations from the same time period include accounts of precoital masturbation by Onan and the contraceptive use of the mandrake root by Leah and Rachel, the wives of Jacob. Sponges, douches, crude methods to stimulate miscarriage, abortifacient methods and techniques—all are described in the classical writings of antiquity. They are alluded to by the philosophers Plato, Aristotle, and Pliny the Elder, as well as by the physicians Hippocrates (460?–?377 B.C.) and Soranus of Ephesus (A.D. 98–138) Soranus clearly distinguished between contraception and abortion, and compiled a remarkably detailed catalogue of methods for each.[1]

This distinction between contraception and abortion is fundamental. It was grasped by Soranus and others long before the study of the human body refined our knowledge of human reproduction. Birth control can be accomplished by either method, contraception or abortion. Contraception (from Latin *contra* 'against') involves the prevention of fertilization or implantation, whereas abortion interferes with the product of fertilization and its complete development during pregnancy.

The Medical History of Contraception, by Himes (1963) is a classic historical study of fertility regulation. The book emphasizes what is perhaps the most important aspect of contraception, namely, motivation. If contraception is to be successful, we need to pay attention to all the factors involved; the person, her motivation, the circumstances of her life and the methods that are available to her.

Theoretical Approaches

Some theoretical approaches to the problem of contraception include interference with the formation and release of the sperm and ovum, preventing their union if they are released, and the prevention of implantation, once the zygote reaches the endometrium.

Understanding the control mechanisms that regulate our ovulatory cycles (Fig. 4-2) can help us determine how the interference with gamete release, or ovulation, would be accomplished. Late in the last century it was known that progesterone could be used to prevent pregnancy in laboratory animals. We know now that this effect occurs because progesterone prevents the release of luteinizing hormone (LH), which is necessary for ovulation. The earliest applications of this concept were directed against infertility, not fertility, in women. Large doses of progesterone were administered in order to supress ovulation. It was hoped that a rebound ovulation and subsequent fertility would follow in the nontreatment cycles. Natural progesterone was expensive and difficult to obtain. Soon, however, advances in steroid biochemistry led to the development of synthetic compounds that acted like progesterone in some ways. These compounds, called progestogens, are one of the substances used in oral contraceptive steroids (OCS).

From our study and understanding of the mechanisms that

[1]Benedict J. Duffy and M. Jean Wallace, *Biological and Medical Aspects of Contraception* (Notre Dame: University of Notre Dame Press, 1969). Reprinted with permission.

were represented in Figure 4-2, we can also see that compounds that would interfere with the action of the releasing factor (GRF) could also act as contraceptives. This idea is being worked on for practical application at some time in the future.

Conception begins with the union of the sperm and the ovum, which is a possible outcome of sexual intercourse. The complete relinquishment of sexual activity is, of course, a thoroughly effective method of contraception. We may know individuals for whom *celibacy* is an accepted professional choice. We may also know others who view it as an onerous burden. Noncelibate couples may also be burdened if the only form of contraception recommended to them involves the restriction of intercourse. This restriction of coitus to a time during the ovulatory cycle when no ova are likely to be present has been called the *rhythm method* of contraception. Various forms of the rhythm method use time as a barrier to the union of sperm and ovum.

Mechanical barrier methods such as the *condom* and *diaphragm* also prevent fertilization. Both methods are more effective when they are used with vaginal creams, foams, and jellies that contain *spermicides.* After *sterilization* in either men or women, gametes and hormones are still formed and released, but they cannot be transported to the place where fertilization occurs.

Conception cannot occur when there is interference with the control mechanisms that prepare the endometrium for implantation. As we have seen, the movement of the zygote down the oviduct to the uterus depends on the proper amounts of estrogen and progesterone. Too much estrogen may hasten this movement, so that the zygote arrives before the endometrium is ready for it. This is the theoretical basis for the administration of large doses of estrogen-like substances in the "morning-after pill." The dangers in this contraceptive method will be considered later in this chapter.

FIGURE 6-1

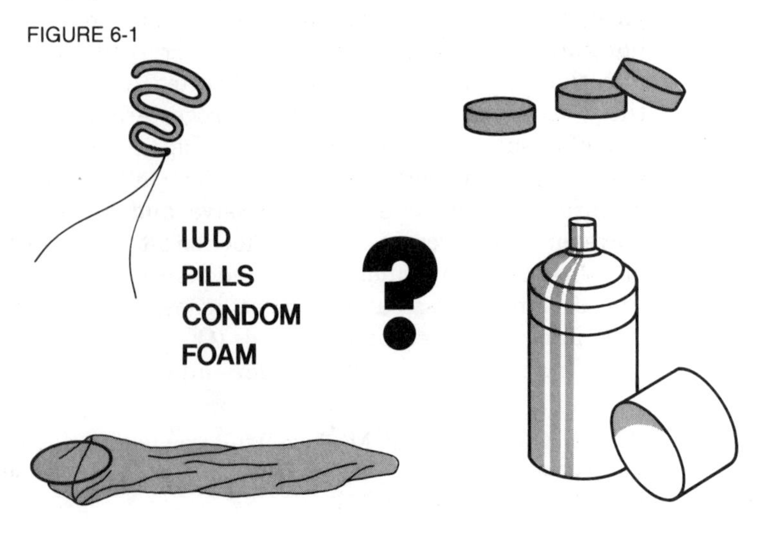

We don't know exactly *how* the intrauterine device (IUD) prevents conception, only *that* it does so. It may prevent implantation by setting up a localized inflammatory response in the endometrium, so that the attachment of the embryo is prevented. It is possible that the presence of the IUD also speeds up the movement of the zygote to the uterus so that it arrives before the endometrium has been hormonally prepared for its implantation.

We may surmise, and correctly, that present-day theories of contraception are based on old ideas. With the exception of the OCS, they are refinements of old barrier and intrauterine methods. These methods have been adapted and packaged to give us a choice when we are motivated to use some kind of contraceptive (Fig. 6-1).

Practical Approaches to Contraception: A Description of Present-Day Methods

ORAL CONTRACEPTIVE STEROIDS The oral contraceptive steroids, as we have seen, prevent ovulation during the normal reproductive cycle. When the OCS pills are taken as prescribed, their contraceptive efficiency cannot be questioned. Besides preventing ovulation, some evidence shows that they also prevent implantation by altering the endometrium. They may also affect the consistency of cervical mucus, so that the sperm are prevented from passing beyond the cervix and on into the uterus and oviduct.

Although we may speak of "the Pill" as if it were one medication, like aspirin, different forms of OCS are available from the various manufacturers. The large drug companies that make OCS use different synthetic forms of estrogens and progestogens and combine them in different amounts and proportions. Since the introduction of the OCS in the 1950s, the dosages of the compounds have been considerably reduced. Table 6-1 shows some of these changes in the dose of steroids per pill that have taken place during the last 20 years. Some side effects of the early OCS may have been related to the larger amounts of hormones in the preparations. As data regarding contraceptive effectiveness were accumulated, it became

TABLE 6–1 Steroid Dosage Changes in OCS

Year	*Amount (mg) of Progestogen (Norethynodrel)*	*Amount (mg) of Estrogen (Mestranol)*
1959	10.00	.150
1969	9.85	.150
	5.00	.075
	2.50	.100
1978	1.00	.050

clear that the dosage could be decreased without sacrificing contraceptive safety. Figure 6-2 illustrates this principle in a dose-response curve. There is a dosage (termed threshold dose) below which no response (in this case, protection against pregnancy), will be observed. Another dose can be determined, that is, the maximum dose, beyond which no additional protection is given. Within the range of the threshold and maximal dose, the proper amount of each hormone, per pill, can be more accurately determined.

Most women who use OCS take pills that combine estrogens and progestogens. These pills are taken every day for 21 days of the pill cycle. The contraceptive activity (inhibition of LH release) is attributed chiefly to the progestogen. The incidence of some undesirable side effects such as breakthrough bleeding is decreased by the addition of small amounts of estrogen. Breakthrough bleeding is vaginal bleeding that occurs unpredictably during the pill cycle. It may be prevented by the estrogen component of the OCS. On the other hand, it is usually the estrogen component that is implicated in the undesirable metabolic effects of the OCS (see Chapter 7).

Contraceptive pills containing only progestogen are available, but they are not widely used at the present time. Pills containing either estrogen or progestogen alone taken sequentially, have been withdrawn from the market (Federal Drug Administration, 1974).

Pill cycles are different from ovulatory cycles in many respects. The vaginal bleeding that occurs after the hormones are withdrawn is predictable. The amount of bleeding may be decreased for those of us who formerly experienced heavy bleeding during the menses. Withdrawal bleeding during a pill cycle occurs because we stop taking the pills, and the endometrium has no more hormonal support. Theoretically, pills could be taken continuously and no withdrawal bleeding would occur. Most of us, however, prefer the reassurance of that bleeding period. We can know when it occurs that we are definitely not pregnant.

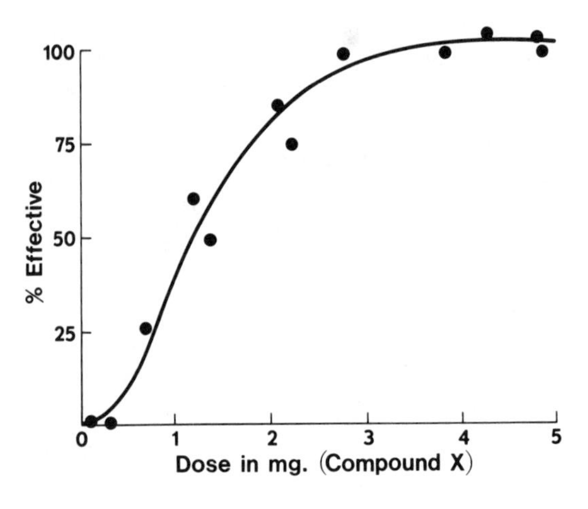

FIGURE 6-2 Theoretical dose-response curve. A dose of about 0.5 mg. of Compound X is the lowest dose at which a response is observed. It is called the threshold dose. The maximum effective dose is about 2.5 mg. Increasing the dosage above this amount does not affect the magnitude of the response.

Some innovations in the hormonal control of conception are being investigated. These include continuous dosages of estrogens and progestogens (Johansson, 1976) and long-lasting and removable vaginal rings which are impregnated with contraceptive steroids (Mishell, 1978). Some of these methods are being tested on women in developing countries. These women may not be able to choose freely whether or not they wish to be experimental subjects in contraceptive research. The Puerto Rican women who were the first subjects for large-scale field trials of the original (high-dose) OCS may also have had no real choice about their participation in these studies. We know that research is needed if better contraceptive methods are to be made available to us. What kind of research? Which women should be subjects? Questions like these, as yet unanswered, are considered in Chapter 7.

Barrier Methods

The Condom. The condom has been a favored contraceptive method for centuries. Prior to the vulcanization of rubber and the development of latex, the condom was made from a variety of inefficient and insecure covering materials. The availability of the modern latex condom has largely solved the problem of security. Present-day condoms are subjected to rigid quality control tests, so that the problem of breaking and perforation are minimized, unless the shelf life (on the label) is exceeded. In order to be effective, the condoms must be put on before *each* instance of intercourse, and removed immediately afterwards. The use of spermicidal creams, foams, or jellies by the female partner increases the effectiveness of the condom.[2] Reactions, favorable and unfavorable, to the condom were summarized in Figures 4-9 and 4-10.

The Diaphragm. The diaphragm is a rubber or latex cup that is inserted into the vagina to cover the cervix. Like the condom, the diaphragm is more effective if spermicidal products are used with it. Steps in the manual insertion of the diaphragm are illustrated in Figure 6-3. Diaphragms must be individually fitted, usually by a physician, nurse, nurse practitioner, or women's health care specialist, who will instruct the user about its correct insertion and removal. After a pregnancy or after considerable weight loss the diaphragm needs to be refitted. The diaphragm must be left in place for a specified time after intercourse.

Vaginal Suppositories. Vaginal suppositories, or inserts, are water-soluble structures with either a gelatin or wax base. They are designed to dissolve at body temperature to release agents that coat the cervix and vagina. The most effective vaginal suppositories con-

[2]Duffy and Wallace, *Contraception.*

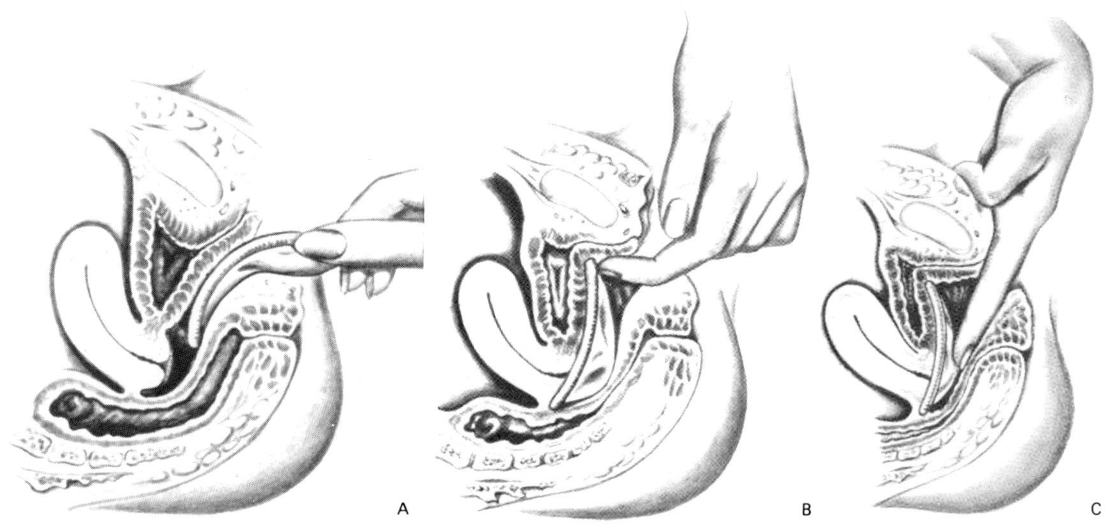

FIGURE 6-3 Steps in the manual insertion of the diaphragm. (A) Compress the diaphragm, insert in vagina and press downward as far as possible. (B) Tip front end upward to fit behind the pubic bone. (C) Check with finger to make certain it fits over the vagina. From a manual for the insertion of the Koromex diaphragm, prepared by the Holland-Rantos Company, Inc., Piscataway, N.J. With permission.

tain spermicides as well as a barrier substance that acts as a temporary chemical diaphragm. For optimal effectiveness, these inserts should be placed in the vagina 10 to 15 minutes prior to intercourse. Spermicidal and barrier type protection may last from 20 minutes to two hours, depending on the product. Vaginal suppositories may have a high failure rate because, unlike foam or cream, they do not coat the cervix uniformly.

Spermicides. The active ingredients of contraceptive creams, foams, jellies, and vaginal suppositories contain chemical compounds which may function as: (1) surface active agents, (2) bacteriocidal agents, or (3) acidic agents. The surface active agents attach themselves to, and disrupt the structure of, the sperm plasma membrane. The selective permeability properties of the cell membrane are affected, resulting in an osmotic disequilibrium that destroys the cell. Surface active agents also interfere with the metabolic processes whereby sperm cells utilize fructose as an energy source. Nonoxynol-9 (nonoxylphenoxypolyoxy ethanol) is a common surface active agent in spermicidal products. Bacteriocidal agents, which may be added to the spermicides, contribute to the effectiveness of the surfactants. Further addition of various acidic agents also enhances spermicidal activity. However, commonly used vinegar and water douches have *no* contraceptive effectiveness.

TIME AS A BARRIER TO CONCEPTION: THE RHYTHM METHOD The contraceptive utilization of this method is based on a very simple physiological principle, namely, that there are only about four or five days during any one ovulatory cycle when intercourse could result in pregnancy. The frustration, confusion, emotional tension, and other negative aspects of this method arise from efforts to determine *which days?* These "unsafe" days are related to the time of ovulation which differs, not only from woman to woman, but also from one cycle to another in each individual woman. Some of us have very regular ovulatory cycles; others must cope with the frustration of never knowing when ovulation or menstruation will occur. Any reliable method of predicting the time of ovulation in women must take these factors into account. The occurrence of ovulation must be known ahead of time in order to allow for sperm survival in the female genital tract. The more exact the method for predicting ovulation, the shorter and more certain the duration of the "unsafe period" during which intercourse should be avoided in order to prevent pregnancy. The same system, in reverse, can be used to optimize the achievement of pregnancy in some infertile marriages. Methods used to calculate the ovulatory period in women include *calendar rhythm,* the *basal body temperature* method, and the *cervical mucus* method.[3]

Calendar Rhythm Method. The earliest and least precise rhythm method relied on mathematics and the calendar. Proposed simultaneously in Japan and Germany, it has been called the Ogino-Knaus method for its originators. Because it is no longer used or recommended, this method will be described only by its chief disadvantage; the very long periods of sexual abstinence required to prevent pregnancy. The only way to determine the fertile period was by a retrospective study of one year's menstrual cycle record, based upon which a day of ovulation could be assumed and an unsafe period calculated. For example, based on the formula developed by Ogino and Knaus, a woman with very irregular cycles might need to mark a period of sexual abstinence that occupied most of her cycle, and yet she could still become pregnant if anxiety, illness, or unusual excitement produced extreme alterations in her cycle length.

More practical rhythm methods rely on indirect measurements of cyclical changes in the ovarian hormones, estrogen and progesterone. These hormones are present in very small amounts in the blood. Their metabolic by-products are also present in the urine. At the present time it is not practical to define the estrogen peak that precedes ovulation (Fig. 4-6) by means of a simple home test. Cyclical changes in either the basal body temperature[4] or the cervical mucus, on the other hand, are simple to measure and/or observe.[5]

[3]Duffy and Wallace, *Contraception.*

[4]Basal body temperature is the temperature recorded in the morning before arising, eating, or engaging in any physical activity.

[5]Duffy and Wallace, *Contraception.*

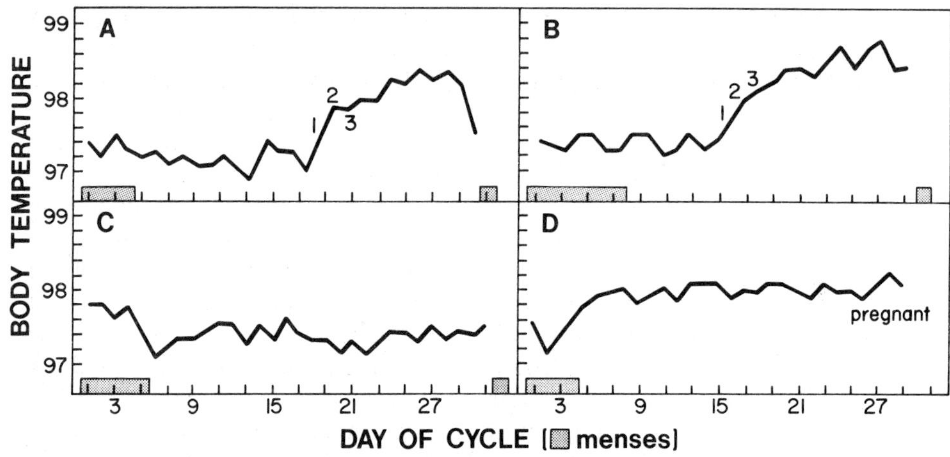

FIGURE 6-4 Basal body temperature (BBT) curves. (A) Typical temperature curve for a normal ovulatory menstrual cycle in which the temperature rise is fast and clear. The high temperature is sustained until the next menstrual period. (B) Temperature curve from normal ovulatory menstrual cycle in which the temperature rise is slow and gradual. (C) Temperature curve of an anovulatory cycle. (D) Temperature curve of a short cycle with early ovulation. Intercourse during menstruation resulted in pregnancy. Adapted from Duffy and Wallace, *Biological and Medical Aspects of Contraception.*

Basal Body Temperature Method. Estrogen has a thermodepressant effect on the basal body temperature. Progesterone produces the opposite effect. Because changes occur in fractions of degrees, thermometers have been especially designed to measure them.

When the basal body temperature is recorded each day on graph paper, a typical diphasic curve may result, as shown on graphs A and B in Figure 6-4. The basal body temperature curve, viewed retrospectively, can demonstrate that ovulation has or has not occurred. Complete protection against pregnancy can only be assured if no intercourse takes place until three days after the rise in the basal body temperature. These days are marked 1, 2, and 3 on graphs A and B. Anovulatory cycles, illustrated in graph C of Figure 6-4 are, of course, safe for intercourse at any time. This can only be determined at the end of the cycle. Ovulation can occur, though rarely, during the bleeding phase of a very short cycle, as illustrated in graph D. This same chart also illustrates the sustained temperature elevation (in the absence of menses) which is usually an indication of pregnancy.[6]

Cervical Mucus Method. The consistency of cervical mucus is also affected by circulating levels of estrogen and progesterone. Around mid-cycle, the cervical mucus that passes into the vagina becomes thick and slippery. Its texture is similar to that of raw egg white and the alignment of molecules within mid-cycle mucus favors move-

[6]Duffy and Wallace, *Contraception.*

ment of sperm through the vagina and cervix. Later in the cycle, under the influence of progesterone, the cervical mucus becomes much thinner. The mucus at this time does not facilitate sperm transport. Laboratory tests for *Spinnbarkeit* (threadiness) and ferning have been used to correlate cervical mucus changes to specific stages in the ovulatory cycle. After a woman has been alerted to observe changes in herself, cervical mucus changes can be used prospectively without laboratory diagnostics to *predict* ovulation before it actually occurs.

The combination of temperature and cervical mucus methods may be used together to pinpoint the time of ovulation in each cycle. Better, more objective methods for determining the time of ovulation await further investigation by reproductive physiologists. Supplementary subjective methods include (1) slight intermenstrual vaginal bleeding, and (2) abdominal pain during the ovulatory period (*Mittelschmerz*). Not all of us, however, experience these latter two symptoms, even when we try to recognize them.

STERILIZATION Production and release of gametes continues after sterilization in both men and women. The union of gametes (fertilization) is prevented by surgical disruption of the ducts concerned with their transport; the vas deferens and oviducts.

In the male, vasectomy involves removal of a segment of the vas deferens. The volume of semen, minus the sperm, is not affected, since the glands associated with the male reproductive tract continue to produce their secretions and empty them into the ejaculatory duct. No effects on erection or ejaculatory response have been reported. Side effects are usually minor and easily correctable.

Sterilization of women via tubal ligation ("tying off the tubes"), may be accomplished by several different methods. The traditional large abdominal incision for sterilization (laparatomy) has been replaced by a tiny abdominal incision (minilaparatomy) or puncture (laparoscopy). Transvaginal approaches to the oviducts are termed culdoscopy or colpotomy. Unless some other reason makes their removal necessary, the ovaries are left intact. Since ovarian hormones travel in the bloodstream, their effects will be the same as before the operation. We will have normal reproductive cycles; however, the ovum cannot be fertilized.

Surgical techniques for performing tubal ligation vary. Sometimes the oviducts are simply cut and tied (ligated). The ducts can also be severed chemically or by burning them. The possibility that the operation can be reversed may be increased as newer surgical techniques become established. At the present time we should assume that the operation is irreversible and that we are choosing a permanent contraceptive method.

THE INTRAUTERINE DEVICE Placing objects within the uterus to prevent pregnancy is an old method of birth control. The ancient

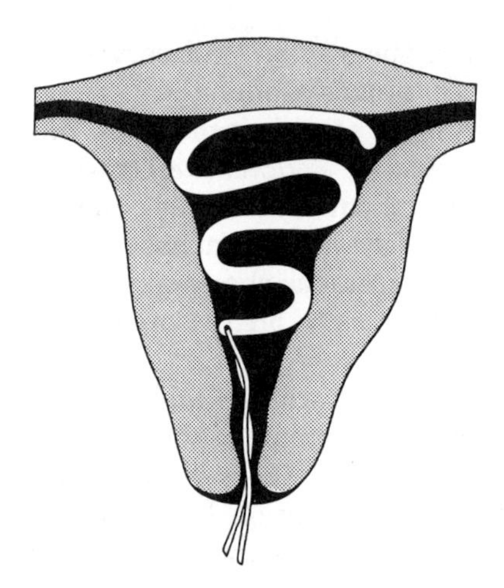

FIGURE 6-5 An intrauterine device (IUD) in place in the uterus.

Greeks were aware of the fact that solid materials introduced into the cervix or uterus had some antifertility effects. Arab tribes have, for many years, prevented pregnancy in camels by inserting stones into their uteri prior to long desert trips.[7]

The first well-documented use of an intrauterine device (IUD) in women was a clinical report in 1909 concerning the contraceptive effect of catgut loops inserted in the cervix. Somewhat later, silver rings, named after their originator, Dr. Graafenberg, were used. Initially, the use of the IUD was strenuously opposed by many physicians. They argued that the placement of a foreign object into a body organ violated fundamental laws of physiological well-being. This negative attitude was reinforced by reports of bleeding episodes, uterine perforations, and pelvic infections. In the absence of effective antibiotic treatment for infection, the use of the IUD was generally condemned in the standard obstetrics and gynecology texts used before 1940.[8]

Advances in plastic technology after the 1940s led to a revived interest in the IUD. The insertion of these devices was simplified by the characteristics of polyethylene plastic, which can be formed, pulled out straight, and fitted into a tube for insertion and then "remembers" and returns to its former shape—loop, coil, ring, or spiral—after insertion. Figure 6-5 illustrates the placement of the IUD in the uterus. The consequences of pelvic infections, if they occur as a result of IUD use, are usually not serious because antibiotics can be prescribed.[9]

Experiments on laboratory animals, including monkeys (primates whose ovulatory cycle resembles that of our species) suggest

[7]Duffy and Wallace, *Contraception.*
[8]Duffy and Wallace, *Contraception.*
[9]Duffy and Wallace, *Contraception.*

that the IUD prevents implantation of a fertilized ovum. It is possible that the presence of the device speeds up ovum transport so that the ovum arrives before the endometrium is prepared to receive it. Alternatively, a localized inflammatory response within the uterus may prevent implantation of the zygote.

Recently, a sensitive radioimmunoassay for HCG (human chorionic gonadotrophin) has been used to test serum levels of this pregnancy hormone in IUD users. The essentially negative results of the test indicated that implantation does not occur in these subjects (Sharpe et al., 1977). In the absence of conclusive evidence or of scientific data to the contrary, it would appear that the IUD acts contraceptively by preventing implantation. The impregnation of IUDs with copper or progestogens seems to promote their contraceptive efficiency.

In recent years the size of the IUD-inserting devices has been modified so that more women can use the IUD. The best time for IUD insertion is after a normal delivery or after an abortion, when the cervix is dilated. In nonpregnancy cycles, the last few days of menstrual bleeding are preferred for IUD insertion. At this time the chances that the IUD will be inserted into a pregnant uterus are minimal.

Postcoital Contraception Large amounts of synthetic estrogens may also prevent the implantation of the zygote. These "morning after pills" may contain diethylstilbestrol (DES). This compound has been associated with cancer in the cervix and vagina (see Chapter 8). If it is administered later than 48 hours after intercourse, the fetus may survive, with possible birth defects. Diethylstilbestrol must be given in large amounts, which may cause extreme nausea unless antiemetics are given at the same time. We can't view the morning-after pill as a miraculous panacea for our contraceptive carelessness. At the present time there are some real dangers associated with its use. We are also not certain exactly how these agents produce their effects. Some animals studies have indicated that DES and similar compounds interfere with implantation, but it is not always possible to say that what happens to laboratory animals happens in the same way to us. The hormonal control systems for ovulation and implantation are different.

If a truly usable form of postcoital contraception is to be made available to us, we need more data to clearly indicate:

1. *whether or not postcoital compounds change the rate of speed with which the fertilized ovum moves through the oviduct*
2. *the possible inhibitory effects of estrogens on the formation of the corpus luteum*
3. *how the estrogens affect the postovulatory development of the endometrium.*

Evaluation of Present-Day Contraceptives

Effectiveness Contraceptive effectiveness can be determined through retrospective studies, which are based on recorded effectiveness in the past. Generally, large populations of women are followed over long periods of time in these studies. The effectiveness of the particular contraceptive method used is expressed in terms of the Pearl formula as the number of pregnancies that occur per 100 women using a particular method of contraception for one year.

Some retrospective studies divide the pregnancies that occur when the method is being used (failures) into method failures and user failures. This is an attempt to differentiate between the correct and incorrect use of the method. Each method has its minimum requirements; for example, the OCS must be taken as prescribed, the diaphragm must be used with every coital act and left in place afterwards, and the IUD must be checked periodically by the user to ascertain that it is still in place.

The contraceptive effectiveness of a particular method, in comparison to other methods, can be judged by examining the data in Table 6-2. These are method failure rates only. Actual failure rates may be higher if these methods are used incorrectly. For example, according to the actual failure rate, the effectiveness of the rhythm method has been reported as between 1 and 47. The Pearl formula for no contraceptive use is between 80 and 90.

Side Effects Physical side effects of oral contraceptive steroids (OCS) and IUDs have been investigated extensively. Other methods are, comparatively speaking, free from side effects.

Oral contraceptive steroids combine social benefits with a high degree of effectiveness. Repeated reports, beginning in Puerto Rico in 1956, proved that various dosages and combinations of OCS furnished sustained fertility control. Lower dosages currently prescribed have eliminated many side effects.[10]

The use of OCS must be discouraged in women at high risk, that is, women over age 40, particularly if they smoke cigarettes, and those with histories of cardiovascular complications, diabetes, liver disease, or nephritis. Patient selection and physician supervision are essential. There is no medical precedent for millions of otherwise healthy, relatively young females taking hormonal agents which have systemwide effects for long periods of time. Women who use OCS should study the patient insert that accompanies the pills. We should also request from our physicians the more detailed FDA brochure on the OCS, or the brochure prepared by the American College of Obstetricians and Gynecologists.

The IUD is a cheap, relatively simple technique of great po-

[10]Duffy and Wallace, *Contraception.*

TABLE 6–2 Failure Rates and Availability of Widely Used Methods of Contraception

Method	*Failure Rate*	*Availability*
Sterilization (male or female)	0.07–0.08	Physician; outpatient or inpatient
Oral contraceptives (combined)	0.14–0.44	Physician; prescription
Oral contraceptives (progesterone only)	3.0	Physician; prescription
IUD	1.1–4.3	Physician insertion
Diaphragm and jelly	2.0	Physician fitting
Diaphragm alone	2.4	Physician fitting
Condom	4.3	Over the counter
Vaginal creams, foams, jellies	14.8	Over the counter

tential usefulness to women who are unable or unwilling to use one of the other methods. It shares with the OCS the advantage of use unrelated to coital activity, and demands even less sustained motivation. Women who are able to retain the device have a high degree of protection; in some women, however, expulsion may occur spontaneously. Spotting and intramenstrual bleeding may occur. Also, as Table 6-2 indicates, the failure rate of the IUD is higher than that of the OCS. Because a small but possible danger of uterine perforation exists, women fitted with IUDs should be examined by physicians at regular intervals.

GOVERNMENT EVALUATION The U.S. Department of Health, Education and Welfare (1978) has evaluated the various methods of contraception as to their effectiveness, advantages and disadvantages, and related health factors. The DHEW findings are summarized below.

STERILIZATION—FEMALE

Effectiveness: *Virtually 100 percent. Pearl formula = 0.08.*

Advantages: *A one-time procedure, permanently contraceptive. Sexual ability and enjoyment not affected.*

Disadvantages: *Surgery required. The procedure should be considered irreversible.*

Health factors: *Surgery related; associated with the general health of the patient.*

STERILIZATION—MALE

Effectiveness: *Virtually 100 percent. Pearl formula = 0.08.*

Advantages: *A one-time procedure, permanently contraceptive. Does not require hospitalization. Studies show that a man's health and sexual ability are not affected.*

Disadvantages: *A man is not completely sterile after the operation. During the ensuing few months another contraceptive must be used. Procedure is considered irreversible.*

Health factors: *Complications occur in 2–9 percent of cases. These include infection, hematomas, granulomas (an inflammatory reaction to sperm absorbed by the body), and swelling and tenderness near the testes. These complications usually do not require surgery.*

THE PILL (OCS)

Effectiveness: *Pearl formula = 0.01–1.0.*

Advantages: *Effective, non–intercourse related.*

Disadvantages: *Must be taken exactly according to prescription.*

Health factors:
1. Pill users have a greater risk than nonusers of having gall bladder disease requiring surgery.
2. Women who smoke should use other forms of contraception.
3. Pill users have a greater risk than nonusers of stroke, blood clotting, and heart attack than nonusers. The risk increases with age. Women over 40 should use other forms of contraception.
4. Other women who should not use the OCS are those who have had *a heart attack, stroke, angina pectoris, or blood clots. Also, those women whose periods are scant or irregular or who have had cancer of the breast or uterus should use other forms of contraception.*
5. Nursing mothers and women who believe they may be pregnant should not use OCS.
6. Use of the OCS accentuates *already existing physiological disorders. These include migraine headaches, mental depression, fibroids of the uterus, heart or kidney disease, asthma, high blood pressure, diabetes, and epilepsy.*
7. If a woman who is taking the OCS desires to become pregnant, she should wait a short time after she stops taking them to conceive. There may be a delay in her ability to conceive, even when no other contraceptive methods are used.
8. Side effects of a relatively minor nature may include tender breasts, nausea or vomiting, gain or loss of weight, breakthrough bleeding, and higher levels of sugar or fat in the blood (see Chapter 7).

INTRAUTERINE DEVICE (IUD)

Effectiveness: *Pearl formula = 1–6.*

Advantages: *No further care needed after insertion. Woman can check placement. Non–intercourse related.*

Disadvantages: *Must be inserted by physician and checked once a year. Some need replacement once every year, some every three years. May cause pain or discomfort when inserted. Cramps and heavier menstrual flow may follow insertion. Device can be expelled without the patient being aware of it, leaving her unprotected. Pregnancy rates are higher than among OCS users, and IUD pregnancies are dangerous.*

Health factors:
1. All of the following are contraindications to use of IUD: abnormalities (including cancer) of the cervix or uterus, heavy menstrual flow or intermenstrual bleeding, severe menstrual cramps, infections of the reproductive tract, uterine surgery, venereal disease, or abnormal Pap smear.
2. Pelvic infection in some IUD users may result in permanent sterility.
3. Before requesting an IUD, the patient should inform her physician about any unexplained genital bleeding or vaginal discharge, allergy to copper, fainting attacks, anemia, prior IUD use, recent pregnancy, abortion, or miscarriage.
4. Serious complications follow if a woman becomes pregnant while wearing an IUD. An IUD wearer who believes she is pregnant should consult her physician immediately, and the device should be removed while the string is still accessible. The IUD gives no protection against extrauterine pregnancies.
5. If the IUD pierces the uterine wall when it is being inserted, surgery is required to remove it.

DIAPHRAGM (WITH CREAM, JELLY OR FOAM)

Effectiveness: *Depends on correct use. Pearl formula = 2–20.*

Advantages: *No routine schedule, no discomfort or cramping, no serious side effects, no effect on physical or chemical body processes.*

Disadvantages: *Must be inserted each time before intercourse. Must be left in place for six hours after intercourse. Some women do not like to touch their bodies in the genital area, and thus will find it a burden to insert the diaphragm. Must be fitted by a physician or women's health care specialist and be refitted if the women gains or loses ten pounds, as well as after childbirth or abortion. A woman whose vagina is greatly relaxed or in whom the uterus has "fallen" cannot use the diaphragm.*

Health factors: *None.*

FOAM, CREAM, OR JELLY

Effectiveness: *Depends on correct use. Pearl formula = 2–36.*

Advantages: *No prescription required. Easy to obtain and use. No serious side effects.*

Disadvantages: *Must be used one hour or less before intercourse. Slight allergic reaction may occur. Relatively high pregnancy rate.*

Health factors: *None.*

VAGINAL SUPPOSITORIES

Effectiveness: *Highly variable.*

Advantages: *No devices needed. No prescription needed. Easy to obtain and use. No adverse side effects.*

Disadvantages: *Intercourse related. Must pay strict attention to time limitations.*

Health factors: *None.*

CONDOM

Effectiveness: *Depends on correct use. Pearl formula = 3–36.*

Advantages: *Cheap, easily available. Gives excellent protection against venereal disease if used consistently. No serious effects.*

Disadvantages: *Intercourse related. Love making must be interrupted before the man enters the woman. Some persons believe that condoms decrease the pleasure in the sex act. The condom can tear or slip during removal from the vagina.*

Health factors: *None.*

RHYTHM METHODS

Effectiveness: *Depends on correct use of method. Pearl formula for calendar method alone = 14–97; for temperature method = 1–20; for cervical mucus method = 1–25; for temperature and mucus = 1–7.*

Advantages: *No drugs or devices needed.*

Disadvantages: *Requires recordkeeping. Instruction needed in use of method. Motivation required to "observe the rules" is greater than for other methods. More difficult to use if menstrual cycles are irregular. Some strain in the husband-wife relationship may occur if the period during which coitus must be avoided is too long. (Another contraceptive could be used during this period.)*

Health factors: *None.*

WITHDRAWAL (COITUS INTERRUPTUS)

The failure rate of this method is so high that it should not be considered as a contraceptive method.

DOUCHING

Use of vaginal douche immediately after intercourse to wash out or inactivate sperm is completely ineffective as a contraceptive method.

Is There an Ideal Contraceptive Agent?

We can make some basic generalizations about the contraceptives that are now available to us:

1. *The most completely effective methods, male and female sterilization, are essentially irreversible.*
2. *The more effective the reversible method, the more serious are the health-related factors.*
3. *Simple methods, with few side effects, are less effective and require more sustained motivation.*

An ideal contraceptive would be effective, safe, reversible, cheap, easy to dispense, and free from physiological and psychological side effects. At present, no such ideal method exists. It seems more and more likely that only a variety of different approaches will come close to providing whatever contraceptive safety is desired in the individual case. Some of us are quite able to use a method that others reject because we have different physiological and psychological reaction patterns. Hopefully, our individual contraceptive needs will be met more satisfactorily by successful research efforts in the future.

Research Prospects for the Future

In spite of impressive advances in the study of human reproductive physiology, more basic research is needed. Unfortunately, the "Golden Age of Grants" of the 1960s is long since over. Federal budgets for research in this area, and in most other areas have been stringently curtailed in recent years. Resources of private foundations have not usually been able to provide adequate funding. In the meantime, some advances are being made, particularly in the areas described below.

Sterilization. The thrust of current efforts is toward the development of methods of sterilization that do not require hospitalization. Options for reversibility are being built into the newer occlusive techniques.

Hormonal Control of Conception. Research on hormonal control of conception is focused on two areas—hormone control at the gonad level and hormone control at the hypothalamic level. In the first area, efforts are being directed toward the development of a male contraceptive similar to the OCS. Serious attention is being paid, also, to the specific adverse side effects of the OCS used by women. In the second area the Nobel Prize in Medicine was awarded in

1977 to investigators who isolated the hypothalamic gonad-releasing factors. It is possible that new contraceptive methods may act at this level through the pituitary gland to produce temporary sterility in the ovaries and testes.

Prediction of Ovulation. Some attention has recently been given to the chemical and physical properties of cervical mucus and to the changes that occur in these properties prior to ovulation. Other compounds in the body also change under the cyclical influence of the ovarian hormones. Many of us might find it useful to have a reliable method of predicting ovulation ahead of time.

Immunization against Pregnancy. The immune response is one of the basic defense mechanisms of the body. This response exquisitely differentiates between "self" and "non-self" and thus combats the invasion of foreign proteins (antigens). The newly formed protective proteins, called antibodies, react only with the antigen that induced their formation. The specificity of the antigen-antibody response, which is diagrammed in Figure 6-6, has been the basis for applications in the field of immunoreproduction. One of the simpler approaches to fertility control through the use of antibodies was the attempt to produce sterility by immunizing the female against spermatozoa. So far, however, this line of approach has been unsuccessful. Sperm contain many different antigens, and it has not been possible to develop an antisperm antibody that would meet the following criteria:

a. form antigen-antibody complexes with sperm only

b. not cross-react with other tissues

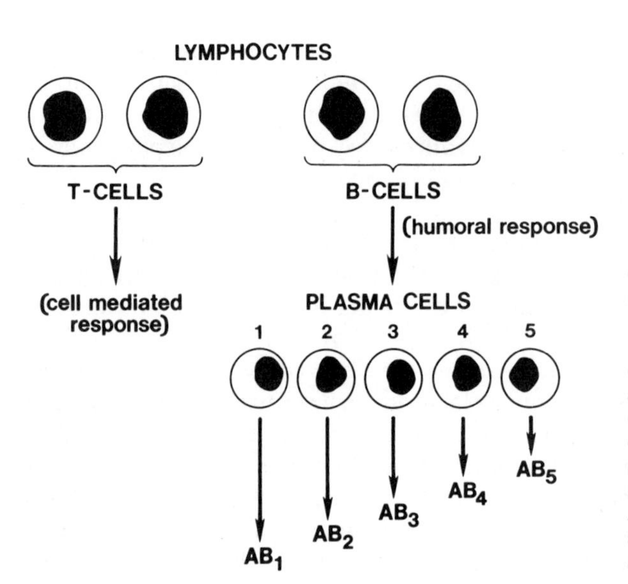

FIGURE 6-6 Fundamentals of the immune response. A foreign antigen stimulates a response from two types of lymphocytes, T-cells and B-cells. T-cells direct cell-mediated immunity. B-cells differentiate into plasma cells which produce an antibody (AB_1, AB_2, AB_3, etc.) which is specific for a particular antigen (1, 2, 3, etc.).

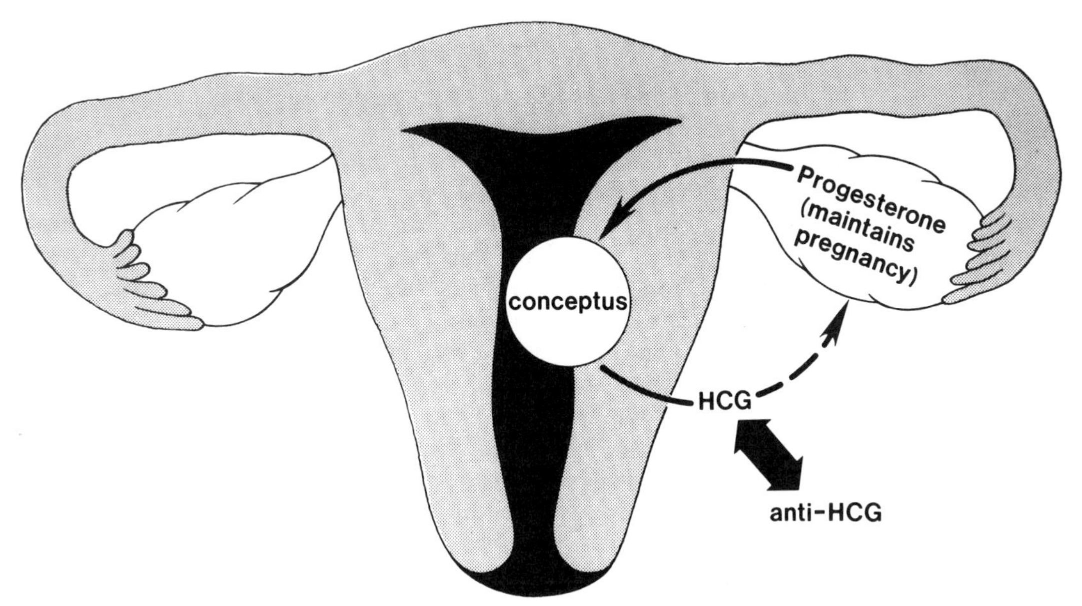

FIGURE 6-7 The immunologic control of conception: Antiserum against HCG. A specific antiserum for the B unit of HCG (anti-HCG) can bind HCG so that it cannot stimulate the ovary to produce progesterone. Pregnancy cannot be maintained without progesterone.

c. *provide sustained and reversible protection against pregnancy in the human.*

Any antibody developed would have to be nontoxic, and must not produce hypersensitivity after continuous administration.

The most specific antigen-antibody reaction under consideration at present involves human chorionic gonadotropin (HCG), the pregnancy hormone discussed in Chapter 5. This protein hormone is composed of two different chains, the alpha chain and the beta chain. Because the alpha chain is identical in structure to the alpha chain of LH and FSH, antisera against the alpha would cross-react among the three hormones. On the other hand, an antibody against the beta chain would interfere with the activity of HCG only. A possible contraceptive activity of anti–beta HCG antiserum is illustrated in Figure 6-7.

Conception and contraception are very important and very personal aspects of our lives. We need the advice and support of a physician whom we trust. Our confidence in our physician's advice will give us peace of mind in following whatever recommendations are made. Very few of us can afford the luxury of impulse buying—we shop around until we are satisfied that we are getting the most for our money. Yet we often hesitate to apply these principles to our selection of a gynecologist. We don't have to continue seeing the first one we visit if we aren't satisfied with his or her treatment of our problems.

While some women are struggling to prevent or delay pregnancy, others are struggling—without success—to achieve pregnancy. Couples who cannot have children when they want them may comprise about 10 percent of our population. What do they do? What would we do if we wanted to have children, tried to become pregnant, and could not? It would be important to locate the part or parts of the faulty control mechanisms and to see if they could be made to function according to their design.

The reproductive potency of our male partner is based on his ability to ejaculate semen that contains adequate numbers of normally shaped, moving sperm. An analysis of his semen by a urologist will indicate whether or not the infertility problem is his.

In our reproductive cycles and systems the causes for infertility may be more complex. The failure to ovulate, for example, may be caused by any of the following conditions:

1. *inadequate releasing factor produced in the hypothalamus or the inhibition of its release*
2. *inadequate production of pituitary gonadotropins or the inhibition of their release*
3. *failure of the ovaries to respond to LH or FSH.*

Induction of Ovulation

Women who do not ovulate because of a failure in one of the control mechanisms listed above can be treated with hormones or with hormone-like substances. Clomiphene (Clomid) is a drug that apparently works within the hypothalamus. It seems to cancel out factors that inhibit the releasing-factor component of the fertility cycle.

Ovulation can also be induced by the administration of the gonadotropic hormones, FSH and LH. Human menopausal gonadotropin (HMG) is obtained for this purpose from women who have undergone the menopause. Human chorionic gonadotropin (HCG) comes from the blood serum of pregnant women. It is chemically similar to LH, and produces similar effects. When HMG and HCG are injected in timed, controlled amounts, they induce the growth and maturation of an ovarian follicle and its release from the ovary.

Both clomiphene and gonadotropin treatments have been used successfully to induce ovulation and to establish pregnancies for previously infertile couples. However, this is a difficult path to parenthood. Long series of hormone measurements and injections are carried out, and coital activity must be scheduled and timed according to possible ovulation induction. Hormone-induced ovulation is not without danger. The ovary may be stimulated to produce not one, but many ova, and multiple births may result. The media give us the news of sextuplets and quintuplets more frequently than in the days of the Dionnes because hormonal intervention is increasing the normal probability of these multiple births. More maturing follicles produce more and more estrogen. This excess estrogen increases the risk of blood clotting, and may also lead to increased accumulation of fluids in the abdominal cavity.

The hormonal control of fertility is obviously a specialized area of medical practice. If we are thinking of using this method for becoming parents, we need to be very careful in our selection of a physician. Most fertility specialists will be located in large medical centers and will have national, if not international, reputations. Time, money, and determination on our part will be needed if we are to locate a competent physician with a history of success in the induction of ovulation.

Structural Disorders

Sterility in women can also be a result of structural disorders in the reproductive system. The factors that can prevent a normal pregnancy include:

1. *Endometriosis, a condition in which islets of endometrial tissue are found scattered throughout the abdominal cavity. They may cause immobilization of sexual organs and prevent conception.*
2. *Secretions within the cervical mucus may be harmful to sperm.*
3. *The oviducts may be blocked so that there is no way for the sperm to reach the ovum, or for the fertilized ovum to move into the uterine cavity. Venereal disease, gonorrhea in particular, may cause permanent sterility in this way.*

Women who are infertile for any reason may or may not eventually be able to have children. They have some options, including adoption. Sometimes surgery can correct the obstruction in the oviduct. As a last resort, a woman may join the thousands of volunteers who are waiting to become mothers of test tube babies.

Test Tube Babies

The birth of Louise Brown in England in 1978 showed that it could be done. (At least two other "test tube babies" have been born since then.) Human ova could be obtained from a human mother and fertilized in vitro with sperm from a prospective father. With the problem of sperm capacitation solved, a suitable medium for the fertilized ovum could be more easily assembled.

A woman who volunteers for this procedure must demonstrate that her reproductive control systems, except for the oviduct problem, are operating within normal limits. This can be done by comparing her hormone secretion patterns with those in Figure 4-6. Even though she is ovulating normally, she is given injections of HMG and HCG to induce the maturation of *several* ovarian follicles. Timing is essential; the physician-investigator must remove the preovulatory follicles from the ovary before ovulation, otherwise they will be in the oviduct and inaccessible. He removes the ova from the follicles by a procedure known as laparoscopy. This operation, with an explanation of how it is performed, is illustrated in Figure 6-8. During the operation the physician punctures the ovary and removes the ova from the follicles with a hypodermic needle.

In the meantime, the husband's sperm, which was collected beforehand, is washed, centrifuged, and placed into the capacitating fertilizing medium. This suspension of sperm is introduced into a culture dish (not literally a test tube!), which contains some inert oil to localize the sperm in a small volume. The ova are introduced into the sperm suspension and their development is observed as they proceed through the early developmental stages of cell division. Up to this point, the procedure used to fertilize human ova is similar to that used when the eggs of research animals, such as mice, are fertilized in the laboratory.

The similarities between fertilization in the human and mouse disappear when we consider the psychological aspects of the procedures. Human beings, who are informed that their sexual union is infertile may be elated or profoundly depressed. If they are depressed, or somewhere near depression, they need help in dealing with this fact in their lives. A new kind of life and life-style, without the care and financial involvement of childrearing, may be appealing after the initial disappointment of the diagnosis has been faced. They may choose to adopt a child, although this procedure has difficulties of its own. If they decide to proceed with programs to induce ovulation, they need many psychological support systems. A schedule of hormone tests, hormone injections, more hormone tests, intercourse on schedule, and pregnancy tests is difficult (and expensive). Consider, at the same time, the alternations between hope and despair which may be repeated over and over again with no success, no pregnancy.

These problems and the need for emotional support are intensified when the mother undergoes the tests and treatments which are necessary for in vitro fertilization, or test tube conception. She needs other hormone injections to prepare her endometrium for implantation when the zygote is removed from the culture dish and placed in her uterus. She needs more support while she is waiting for the results of the pregnancy test.

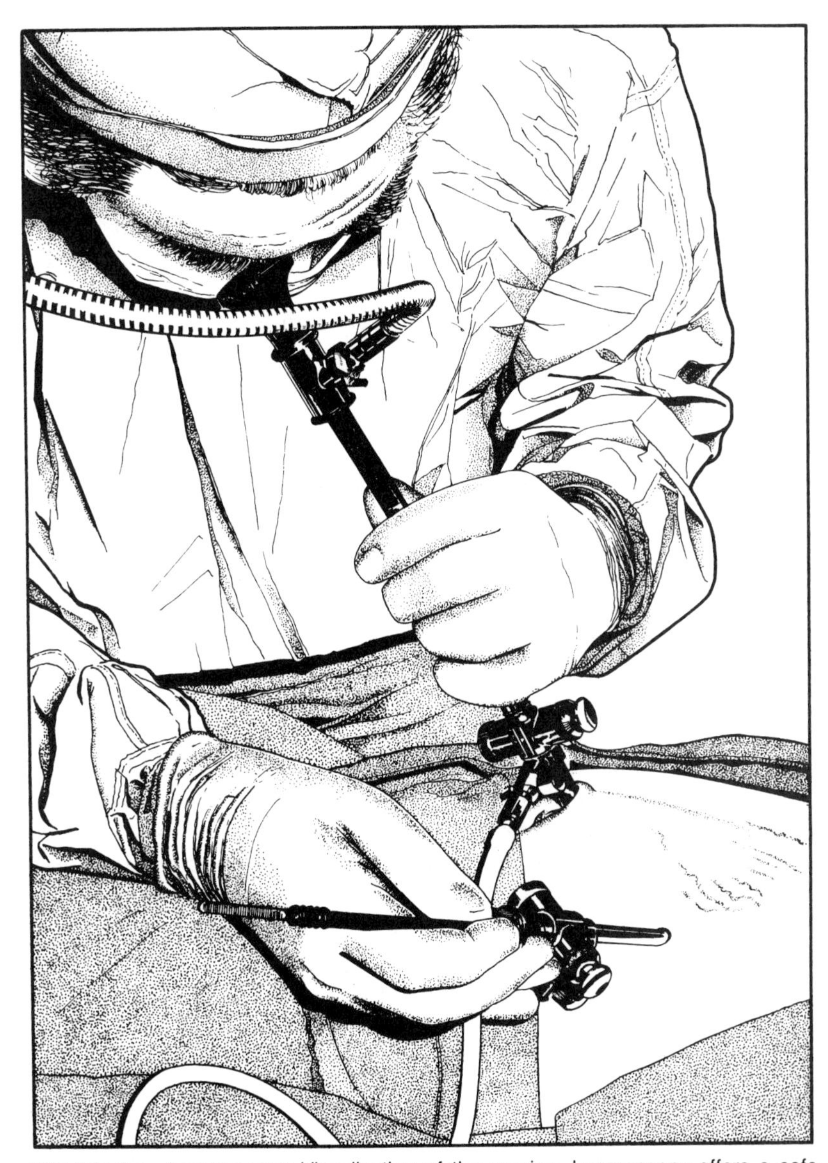

FIGURE 6-8 Laparoscopy: Visualization of the ovaries. Laparoscopy offers a safe and simple method for performing various abdominal operations. Light from an external source is transmitted through glass fibres (enclosed in the lead in the upper part of the illustration) and down the telescope. The surgeon has a clear vision of the abdominal organs. The abdomen is distended with an inert gas to give the surgeon room to manoeuvre his instruments. In this illustration he is using (with his right hand) a pair of specially designed manipulating forceps. Notice the minor degree of damage to the abdominal wall. Laparoscopy can be used for many operations such as tubal sterilization, cauterizing endometriosis, taking biopsies from liver or elsewhere, examining the oviducts, or collecting oocytes and fluids from Graafian follicles. Other forms of endoscopy permit the fetus to be examined, or evaluation of the uterine and oviducal walls. Sometimes the patients can return home on the same day that the operation was performed. From R. G. Edwards, "Control of Human Development," in *Artificial Control of Conception* ed. C. R. Austin and A. V. Short (Cambridge: Cambridge University Press, 1972), p. 94a. With permission.

We don't know how many failures in the method, or in the way it was carried out, preceded and followed the successful attempts at in vitro fertilization that resulted in the birth of healthy, normal children. It is difficult to say when or where the next test tube baby will make its appearance. The in vitro fertilization of the human ovum is one way of controlling human conception. It is not a particularly difficult scientific innovation, but it is a very complicated one. The complications involve ethical, social, and psychological issues that are of concern to all of us. These issues have been virtually unaddressed, particularly in a female forum.

ABORTION

Abortion is the termination of a pregnancy before the fetus is viable, that is, before it can live outside the uterus without the placenta. Usually, this means before the twenty-eighth week of pregnancy, but technological developments are permitting the survival of younger and younger fetuses.

We may think of abortion as something a woman chooses to have done when she doesn't want to have the child. Medically, however, there are several kinds of abortions, some of which occur unintentionally in spite of the best obstetrical care. See Table 6-3.

Induced abortions are those brought about by some external intervention. Elective abortion in the first trimester, that is, within the first 12 weeks of pregnancy, is legal in the United States. A 1973 U.S. Supreme Court decision gave the pregnant woman and her physician the right to terminate early pregnancy. Some individuals and groups have questioned this decision, others have cheered

TABLE 6-3 Kinds of Abortions

Kind	*Cause*
Accidental	Injury, infection, etc.
Complete (or incomplete)	The entire contents of the uterus—fetus, fetal membranes, and placenta—are eliminated (or parts are left behind).
Elective	Induced abortion without a specific medical indication.
Habitual or recurrent	Loss of three or more successive pregnancies.
Induced	Intentionally caused.
Inevitable	Condition characterized by severe vaginal bleeding and uterine contractions. This is accompanied by cervical dilation which has progressed so far that there is no hope of preserving the fetus.
Missed	The fetus dies and remains in the uterus for two months or longer.
Septic	The embryo and maternal organs are infected.
Spontaneous	Termination of pregnancy without apparent cause.
Therapeutic	Termination of pregnancy for medical or psychiatric reasons.
Threatened	Slight vaginal bleeding with or without slight uterine contractions. No cervical dilation.

it. Questioning or cheering, each of us has a right to know what options are open to us when we are pregnant and cannot cope with the consequences, namely, the birth and care of the child.

First Trimester Abortion

First trimester abortions include those performed within the first 12 weeks of gestation, measured from the last menstrual period (LMP). This operation has become a common surgical procedure. During 1975, for example, more than 700,000 legal first trimester abortions were reported to the Center for Disease Control.[11] The surgery involves (1) administration of anesthesia, (2) dilation of the cervix, and (3) removal of the products of conception—the fetus, membranes, and placenta—from the uterus.

Either general or local anesthesia may be used. With general anesthesia the immediate postoperative complications seem to be more common. Blood loss, injury to the cervix, uterine perforations and hemorrhage, and possible abdominal hemorrhage are more likely. The complications, however, occur in fewer than 0.5 percent of all cases. The aftereffects of a local anesthetic are fewer, but the rate of fever afterward is higher than for general anesthesia (Grimes et al., 1979). The pain associated with local anesthesia for first trimester abortion has been rated by over 2,000 women who experienced it. On the average, the women rated the pain less than an earache or a toothache, but more than a headache or a backache. The pain was inversely related to age, and directly related to the fearfulness of the patient. When gestation had proceeded of 7 weeks or less, or 12 weeks or more, patients experienced more pain than when gestation had proceeded 8–11 weeks (Smith et al., 1979).

In order for the contents of the uterus to be removed, the cervix must be dilated (enlarged). The extent of dilation is directly related to gestational age. Very early abortions can be performed seven to eight weeks after the LMP with little or no cervical dilation. On the other hand, the larger size of the fetus later in development makes it necessary to use a tube with a wider diameter. The cervix must be dilated to accommodate this wider tube.

Dilation of the cervix is frequently the most difficult and painful part of the abortion procedure, and it is highly dependent on the physician's skill (Ott, 1977). The sequence of events in cervical dilation may be as follows:

1. *Bimanual pelvic examination by the physician to determine size, shape, and position of the uterus.*

[11]Center for Disease Control, *Abortion Surveillance, 1975.* Washington, D.C.: DHEW (April, 1977).

2. *Insertion of speculum into the vagina to expose the cervix.*
3. *Cleansing of cervix and vagina with antiseptic solution.*
4. *Placement of tenaculum (a holding instrument) on the cervix to hold it in place.*
5. *Administration of local anesthetic if a general anesthetic has not been administered beforehand.*
6. *After the anesthetic takes effect, a sound is passed into the uterus. A sound is a cylindrical, usually curved, metal instrument used for exploring body cavities.*
7. *If there is no problem, dilation begins with the smallest dilator used first, followed by progressively larger sizes until the cervix is dilated just enough to allow the entry of whatever instruments are going to be used (Ott, 1977). See Figure 6-9.*

After the cervix has been dilated to the required diameter, the uterine contents are removed, either by curettage (scraping) or by a vacuum aspiration of the uterus. A dilation and curettage (D and C) is a standard gynecological procedure used to treat infertility, menstrual irregularity, and excessively long bleeding periods. Before the development of the vacuum suction methods, the D and C was the standard method for first trimester abortions. The physician used a curette, which is a metal loop at the end of a long thin handle to loosen the endometrium. The endometrial tissue was removed with a forceps. Most first trimester abortions today use suction aspiration to remove the endometrial tissue (Grimes et al., 1979).

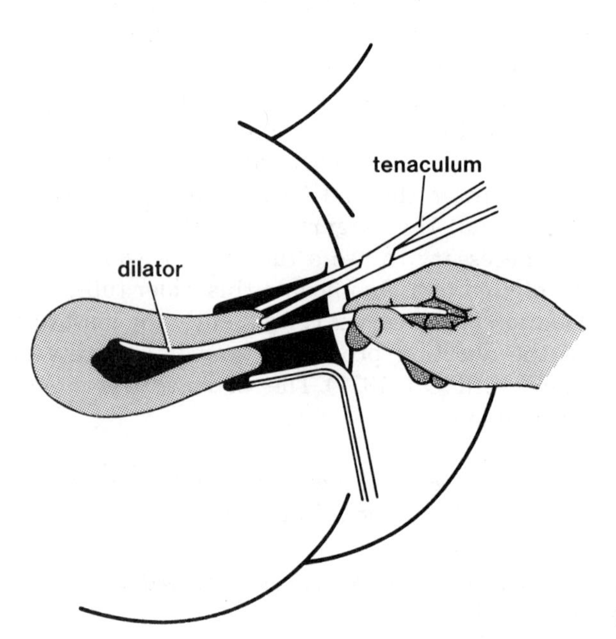

FIGURE 6-9 Cervical Dilation.

First trimester abortions can be done in clinics, rather than in hospitals. What kinds of women seek these abortions? Data from a recent study of over 30,000 women who received first trimester abortions (Grimes, et al., 1979) showed the following age distribution: One-third were below 19, one-third were 20–24, and the remainder were over 25. Over half of these women were white; 22 percent were married. Nearly half had had no prior pregnancies.

Pregnancy Termination at Mid-Trimester

Abortion between 13 and 26 weeks of gestation is a more complicated procedure. It is a hospital procedure, and not all hospitals will allow it to be performed. For these reasons, most women who decide to have an abortion (elective and induced) have it in the first trimester. Reasons for delay may include:

1. *Lack of information or incorrect information about the availability of first trimester abortion.*
2. *Ignorance or psychological denial of pregnancy until signs and symptoms are obvious.*
3. *Indecision regarding the desirability of abortion.*
4. *Late identification of maternal medical problems that contraindicate pregnancy.*
5. *Discovery of fetal abnormalities. These may be detected by amniocentesis, the results of which are not known until early in the second trimester of pregnancy (see Chapter 2).*

Surgical interruption of mid-trimester pregnancy by hysterectomy (removal of the uterus and its contents) or hysterotomy (removal of the fetus, as in a Cesarean section) are not frequently used. The most common method of mid-trimester abortion involves the injection of a fluid into the amniotic sac of the fetus. The fluid contains a substance which, because of its osmotic[12] characteristics, will cause water to move into the amniotic sac until it swells and breaks. The fluid may also contain oxytocin and prostaglandin. These compounds cause uterine contractions during parturition. They have similar effects in causing the premature expulsion of the fetus in a mid-trimester abortion. The time between the injection of the abortifacient substance and the expulsion of the fetus may vary between 12 and

[12]Osmosis is the process by which water moves from its greater to its lesser concentration. If the solution injected into the amnion is hypertonic, it contains fewer water molecules than the solutions outside the amnion. Water will move into the amnion which will burst, initiating premature expulsion of the fetus.

20 or more hours. The placenta is usually delivered within two hours of fetal expulsion (Strauss et al., 1979).

Complications of mid-trimester abortions include failed abortions, blood loss, infection, and cervical lacerations. "The development of a safe, simple, and efficient method of mid-trimester pregnancy termination forms a continuous challenge to fertility control" (Bienariz, 1979).

A woman who is pregnant and cannot afford to raise a child has other options besides elective abortion. These options will vary with her economic and marital status. Both married and unmarried women should consider their options with reference to their own priorities.

REFERENCES

CONTRACEPTION

Chang, M. C. "Development of the Oral Contraceptives," *Am. J. Obstet. Gynecol.* 132, no. 2 (September 15, 1978): 217–19.

Drill, V. A. "History of the First Oral Contraceptive," *J. Toxicol. and Env. Health* 3 (1977): 133–38.

Duffy, B. J., and Wallace, M. J. *Biological and Medical Aspects of Contraception.* Notre Dame: University of Notre Dame Press, 1969.

FDA Drug Bulletin, no. 6 (1976): 26–27.

Gillett, P. G. "Immunologic Control of Fertility: Search for a Contraceptive Vaccine," *Clin. Obstet. Gynecol.* 20, no. 3 (September 1977): 705–15.

Hearn, J. P. "Immunization Against Pregnancy," *Proc. Rov. Soc. Lond. B.* 195 (1976): 149–60.

Himes, Norman E. *The Medical History of Contraception.* New York: Gamut Press, 1963.

Johansson, E. D. B. "Advantages and Disadvantages of the Intrauterine Device and the Hormonal Implant," *Proc. Rov. Soc. Lond. B.* 195 (1976): 81–91.

Lane, Mary E.; Rosalind Arceo; and Aquiles J. Sobrero. "Successful Use of the Diaphragm and Jelly by a Young Population: Report of a Clinical Study," *Family Planning Perspectives* 8, no. 2 (March-April 1976): 81–86.

McCarthy, J. J. "Cervical Mucus: A New Dimension for Family Planning," *J. Fla. Med. Assoc.* 65 (1978): 22–24.

Mishell, Daniel R.; Donald E. Moore; Subir Roy; Paul F. Brenner; and Mary Alice Page. "Clinical Performance and Endocrine Profiles with Contraceptive Rings Containing a Combination of Estradiol and d-Norgestrel," *Am. J. Obstet. Gynecol.* 130, no. 1 (January 1, 1978): 55–62.

Noller, F. K. "The Morning-After Pill," *Fertil. Steril.* 28 (1977): 211.

Odeblad, E. "Physical Properties of Cervical Mucus," in *Cervical Mucus in Human Reproduction*, World Health Organization Colloquim, Geneva, ed. M. Elstein. Copenhagen: Scriptor, 1972.

Porter, J. F. "The Rhythm Method of Contraception," *J. Reprod. Fertil.*, Supp. 22 (1975): 91–106.

Sharpe, R. M.; W. Wrixon; B. M. Hobson; C. S. Corker; H. A. McLean; and R. V. Short. "Absence of HCG-Like Activity in the Blood of Women Fitted with Intrauterine Contraceptive Devices," *J. Clin. Endocr. Metab.* 45 (1977): 496–99.

Toppozada, M. "The Clinical Use of Monthly Injectable Contraceptive Preparations," *Obstet. Gynec. Survey* 32, no. 6 (1977): 335–42.

U.S. Department of Health, Education and Welfare, "Contraception: Comparing the Options," HEW publication FDA 78-3069. Washington, D.C.: DHEW, 1978.

Vessey, M. P., and R. Doll. "Evaluation of Existing Techniques: Is the Pill Safe Enough to Continue Using?" *Proc. Rov. Soc. Lond. B* 195 (1976): 69–80.

Yuzpe, A. A. "Use, Risks, Side Effects Associated with Postcoital Hormonal Contraceptives," *Int. J. Gynaecol. Obstet.* 15 (1977): 133–36.

INFERTILITY

Edwards, R. G. "Control of Human Development," in *Artificial Control of Reproduction,* ed. C. R. Austin and H. V. Short. Cambridge: The University Press, 1972.

Kolata, Gina B. "In Vitro Fertilization: Is It Safe and Repeatable?" *Science* 201 (August 25, 1978): 698–99.

———. "Infertility: Promising New Treatments," *Science* 202 (October 13, 1978): 200–203.

ABORTION

Bieniarz, Joseph. Discussion, *Am. J. Obstet. Gynecol.* 134, no. 3 (June 1, 1979): 264.

Chaudry, Susan L. "Pregnancy Termination in Mid-Trimester: Review of Major Methods," *Population Reports,* Series F, no. 5 (September 1976): F65–F84.

Grimes, David A.; Kenneth F. Shulz; Willard Cates, Jr.; and Carl W. Tyler, Jr. "Local vs. General Anesthesia: Which Is Safer for Performing Suction Curettage Abortions?" *Am. J. Obstet. Gynecol.* 135, no. 8 (December 15, 1979): 1030–35.

Lauersen, Niels H.; Steven Seidman; and Kathleen H. Wilson. "Cervical Priming Prior to First-Trimester Suction Abortion with a Single 15-Methyl-Prostaglandin $F_{2\alpha}$ Vaginal Suppository," *Am. J. Obstet. Gynecol.* 135, no. 8 (December 15, 1979): 1116–18.

Ott, Emiline Royco. "Cervical Dilatation: A Review," *Population Reports,* Series F, no. 6 (September 1977): F85–F104.

Smith, Gene M.; Phillip G. Stubblefield; Linda Chirchirillo; and M. J. McCarthy. "Pain of First-Trimester Abortion: Its Quantification and Relation with Other Variables," *Am. J. Obstet. Gynecol.* 133, no. 5 (March 1, 1979): 489–98.

Strauss, John H.; Miles Wilson; David Caldwell; Warren Otterson; and Alice O. Martin. "Laminaria Use in Midtrimester Abortions Induced by Intra-amniotic Prostaglandin with Urea and Intravenous Oxytocin," *Am. J. Obstet. Gynecol.* 134, no. 3 (June 1, 1979): 260–64.

Zimmerman, Margot. "Abortion Law and Practice: A Status Report," *Population Reports,* Series E, no. 3 (March 1976): E25–E40.

7

Female Hormones: Research Findings and Research Issues

Research into the effects of hormones on our physiology and behavior has many important implications. If we are taking synthetic hormones in the form of oral contraceptive steroids (OCS) or estrogen replacement therapy (ERT), we are concerned to know how these substances affect our physiological functioning and what, if any, health-threatening risks are involved. If we experience extreme fluctuations in mood or behavior patterns during our reproductive cycles, we may be concerned to know if these fluctuations are normal or if they indicate some individual disorder or malfunctioning. Because many research projects have been designed to explore these and other questions about the effects of hormones in individual women and large populations of women, we ourselves may be asked to participate in one or more of these studies. In this chapter, we shall review some research findings already reported on the physiological risks of synthetic hormone use and on the relationship between hormone levels and behavior. We shall also discuss some ways of evaluating the benefits and risks of new research projects in which we ourselves may become involved.

HORMONES, GENES, AND METABOLISM

We have seen how our bodies develop under the influence of the hormones that begin to exert their effects during puberty. The hypothalamic releasing factor, the pituitary gland, and the ovaries interact to control our reproductive cycles, which end with the menopause.

Cyclic fertility can be inhibited by the use of oral contraceptive steroids, synthetic estrogens and progestogens of unquestioned contraceptive effectiveness. It is also possible to alleviate some symptoms of the menopause by estrogen replacement therapy. The use of estrogens and progestogens (OCS) or conjugated estrogens (ERT) is not recommended for all women, however. Why not? The answer to this question may involve our genetic differences.

Our genetic inheritance includes, among many other actual and potential factors, a body chemistry that may have some peculiarities. In other words, our metabolism may not be like everyone—or anyone—else's. What is metabolism? It is, in a broad sense, the sum of all the chemical reactions that maintain our bodies. If we want to take charge of our own lives and health, we need to know something about our own biochemistry, because some of it may be affected by the ovarian hormones or their synthetic counterparts.

All the cells of our bodies have certain common requirements. They need oxygen, energy, a constant temperature, a definite hydrogen ion concentration, a means of waste elimination, and raw materials to reproduce or to repair themselves. Each system of our bodies plays a unique role in the biochemistry of our metabolic processes. For example, our digestive system breaks down large food molecules into smaller ones that can be used by our cells as energy sources. There are other systems, homeostatic control systems, that ensure that all conditions required by our cells are met. See Figure 7-1. When our bodies are in a state of homeostasis the cellular environment is within the limits of efficient functioning.

Chapter 1 outlined the coordinated activities of the control systems that regulate our bodies. How do the estrogens and progestogens in the OCS and the conjugated estrogens used in ERT affect our

FIGURE 7-1 Homeostatic regulatory systems control entrance to and from the cell.

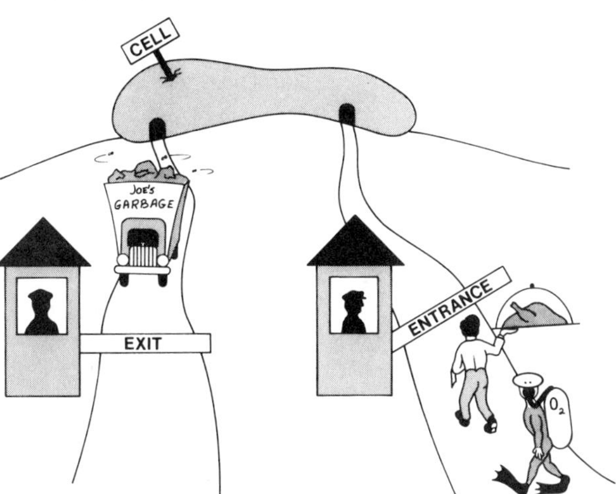

metabolism? That question has been asked before, and many women who take these hormonal medications have become the subjects of research investigations. Large groups of healthy women taking OCS for measurable lengths of time have been used by research scientists to gather information about the effects of OCS on our metabolic functions. The best-studied systems in our bodies, insofar as OCS effects are concerned, have been (1) systems that regulate the utilization of food to produce energy, (2) the cardiovascular system, and (3) the system that regulates kidney function. Although in some cases the data are contradictory, in general, they indicate that some of us, because of inherited differences in our metabolism, should not take OCS. Are we among these women? Let's see.

PHYSIOLOGICAL ACTION OF STEROID HORMONES

In Chapter 1 we looked at the general way in which steroid hormones produce their effects in target cells. Estrogen and progesterone are steroid hormones. They produce their effects in their target cells by passing through the cell membrane and attaching themselves to a receptor molecule. The hormone-receptor complex induces the synthesis of a new protein. The DNA in the nucleus serves as a template, and the amino acid fragments are assembled by RNA. The newly synthesized protein could be a new enzyme capable of changing the normal metabolic activity of our cells.

Very few data are available to show precisely, in terms of this

TABLE 7–1 Systemic Changes during the Menstrual Cycle

System	*Change*	*Increase (↑) or Decrease (↓) with Estrogen (E) or Progesterone (P)*
Thermoregulatory	Basal body temperature	E↓, P↑
Cardiovascular	Blood pressure	E↑, P↓
Blood		
Formed elements		
Red blood cells		E↓
White blood cells		E↓
Platelets		E↑
Plasma and serum		
Carbohydrates	Lowered glucose tolerance	
Lipids and cholesterol		E↓
Amino acids		Variable
Respiratory	CO_2 tension	P↓
Body weight		E↑

Source: Based on data from Anna L. Southam and Florante P. Gonzaga, "Systemic Changes during the Menstrual Cycle," *Am. J. Obstet. Gynecol.* 91, no. 1 (January 1, 1965): 142–165.

TABLE 7–2 Potential Effects of Oral Contraceptives on the Results of Laboratory Tests

System or Function	*Substance Measured*	*Effect*[a]
Nutrition and energy-furnishing processes	Glucose tolerance	Moderate decrease (depends on preparation)
	Insulin	Moderate increase
	Cholesterol	Moderate increase (depends on preparation)
	Lipoproteins (total)	Moderate increase
	Triglycerides	Moderate increase (estrogen)
	Phospholipids (total)	Moderate increase
Water and electrolyte balance	Aldosterone	Moderate increase
	Angiotensin	Moderate increase
	Angiotensinogen	Moderate increase
	Renin	Decreased; renin activity increased
	Sodium	Moderate decrease
Blood clotting	Antithrombin II	Moderate decrease
	Fibrinogen	Slight increase (estrogen)
	Plasmin	Moderate increase
	Plasminogen	Moderate increase
	Platelet aggregation	Moderate increase

[a]Effect is (1) dependent upon type of preparation used, (2) exaggerated in women whose smaller physiological reserves may be challenged.

Source: J.B. Miale and J.W. Kent, "The Effects of Oral Contraceptives on the Result of Laboratory Tests," *Am. J. Obstet. Gynecol.* 120, no.2 (1974): 265, by permission of C.V. Mosby Company and the authors.

theory of hormone action, how estrogen and progesterone bring about changes at the cell level in the human female. However, systemic changes during the menstrual cycle have been noted. The effects of estrogen and progesterone on the basal body temperature were described in Chapter 6. Other systemic effects of estrogen and progesterone during the menstrual cycle are outlined in Table 7-1. Laboratory tests (Table 7-2) indicate that OCS medication causes changes in bodily functions that are concerned with homeostasis: nutrition and energy furnishing processes, water and electrolyte balance, and blood clotting.

The Metabolism of Nutrients

The food we eat is broken down into smaller molecules by our digestive systems. Nutrients we ingest as energy sources can be classified as proteins, fats, and carbohydrates (starches and sugars). See Table 7-3.

TABLE 7–3 Nutrients in Our Bodies

Energy-Yielding Nutrients	*Other Nutrients*
Carbohydrates	Water
Proteins	Vitamins
Fats	Minerals

GLUCOSE UTILIZATION Glucose is a sugar and a potential energy source for our cells. The other sugars can be converted to glucose by a series of enzyme-directed reactions. Glucose, in turn, can be chemically converted to glycogen by another series of reactions. In the form of glycogen, glucose is stored in the muscles and liver. Glycogen can be reconverted to glucose when the body's energy needs require it. Carbohydrates in the form of glucose are carried by the blood to all the cells of the body. If glucose passes through the cell membrane it is broken down by another series of enzyme-facilitated reactions to form still smaller molecules. Energy is given off during these reactions and becomes available to the cell.

Glucose cannot move from the bloodstream across the membrane and enter a cell if insulin is not available. Insulin, a hormone, is secreted by the islets of Langerhans in the pancreas. In normal individuals the stimulus for insulin secretion is a certain concentration of glucose in the blood. When blood glucose reaches a critical level, insulin is secreted and glucose enters the cell where it is either used for immediate energy or converted to glycogen and stored.

Blood glucose and insulin interact with each other in a pattern characteristic of the hormonal negative feedback system. If insulin is not present and glucose cannot enter the cell, it is retained in the bloodstream and gives rise to a condition commonly known as high blood sugar. Glucose is also excreted in the urine, a phenomenon described as passing or spilling sugar. People who suffer from juvenile onset diabetes do not have enough insulin to facilitate the transfer of glucose to their cells. On the other hand, many people who become diabetic as adults have enough insulin in their bloodstream. The transport of glucose into their cells is impaired in another way.

The effective use of glucose in our bodies thus requires adequate insulin production as well as the effective transportation of glucose into our cells. Glucose tolerance tests are clinical measurements of the body's ability to handle glucose. In the test, a measured amount of glucose is given by mouth and blood glucose is measured during a certain period of time after glucose ingestion. Any abnormality in the results of a glucose tolerance test may indicate that the individual's ability to secrete insulin is impaired or that some other factor is interfering with the action of insulin at the cell level.

In order to measure the effects of gonadal steroids on carbohydrate metabolism, oral glucose tolerance tests have been carried out in women during different stages of the menstrual cycle (Cudworth, 1975). No abnormal glucose tolerance tests were noted. Apparently, the hormones that our bodies produce during the normal reproductive cycle do not interfere with glucose metabolism.

This may not be true, however, for the synthetic hormones found in the OCS. Oral contraceptive preparations that contained 0.1 mg of synthetic estrogen and 1 mg of a progestogen affected glucose tolerance in about 17 percent of women who took them for three years (Spellacy, et al., 1977). These women were tested before they began taking OCS, and their glucose tolerance tests were normal at that time. After they stopped taking the OCS medication, their glucose tolerance tests returned to the normal range in most cases.

Apparently the synthetic ovarian steroids in the OCS are responsible for the impairment, in some women, of their ability to respond to the physiological challenge of excess glucose. Although the way in which this happens is not clear, the consequences for women who have diabetes or a family history of diabetes are clear. These women may experience difficulties in metabolizing carbohydrates if they use OCS. Pregnant diabetic women may have similar problems and are usually watched carefully during their pregnancies for adjustments in their insulin dose. In some women, the higher-than-normal hormone levels induced by pregnancy or the use of OCS can reveal a genetic weakness in their ability to metabolize carbohydrates.

AMINO ACIDS Like glucose, amino acids require insulin for entry into our cells where they are used to synthesize new structural or functional proteins or to furnish energy when glucose and fat stores have been used up. We recall that during a normal reproductive cycle, progesterone is present in larger amounts during the luteal phase, which occurs after ovulation in the second half of the cycle. Blood levels of amino acids are lower during this phase of the cycle than during the preovulatory or follicular phase (Cox and Calhane, 1978). It is possible that progesterone augments the action of insulin so that amino acids pass more readily from the blood to the cells.

The effects of OCS on the way our bodies use individual amino acids have also been studied. In some cases, OCS ingestion produces abnormal metabolic patterns for amino acid breakdown, particularly if these amino acids are present in quantities higher than those found in a normal diet (Rose and Adams, 1972).

FATS (LIPIDS) The association between fatty substances such as cholesterol and cardiovascular diseases has motivated serious efforts to determine the effects of OCS and ERT on the way our bodies use fats. Lipid (fat) research centers in the United States, Canada,

Russia, and Israel conducted large population studies between 1971 and 1976 (Wallace et al., 1977). These geographically different centers standardized their procedures for:

1. *Selecting subjects. Only nonpregnant white women who were not receiving medication to change any of the elements being studied were used in the study.*
2. *Blood collection. All blood samples were collected after a 12-hour fast.*
3. *Analyses. The same methods of analyses were used for all the studies.*

More than 16,000 women were included in these studies. After all the facts were obtained, they indicated that, in younger women, increases in plasma lipids such as cholesterol were consistently associated with the use of the OCS. Women with a personal history of abnormal fat metabolism showed even more substantial increases. These increased blood levels of fatty substances, which seem to be a direct effect of OCS use, are particularly dangerous to women who have histories of cardiovascular disease. Fatty substances may accumulate in the blood vessels and lead eventually to the formation of blood clots.

Older women who use estrogens for replacement therapy responded in the opposite way. In most cases the hormone lowered their blood levels of cholesterol and other fats as compared to control subjects in the same age group who do not use ERT.

SUMMARY Evidence from a number of sources suggests that the gonadal hormones we ingest may change the way our bodies metabolize nutrients. Some of these changes are mild and transient. They cause no unfavorable symptoms and we revert to normal when the medication is stopped. On the other hand, serious symptoms may develop if we have any genetically caused irregularity in the way our bodies metabolize nutrients. Women who have a personal or family history of genetic disorders, such as diabetes, should be particularly careful. The mechanisms by which OCS prevent pregnancy are well known; the mechanisms by which they affect the metabolism of nutrients are not well known. They may involve and upset the balance of other endocrine control systems in our bodies.

Cardiovascular and Kidney Function

The heart, blood vessels, and kidneys are important organs in the maintenance of optimal working conditions within our bodies. The proper functioning of the cardiovascular system requires that the pump (heart) and blood vessels (arteries and veins) be whole and healthy. When the cardiovascular system is operating according to its design, food and oxygen are transported to all the cells of our bodies. Carbon dioxide and other waste products of metabolism are removed for elimination by the lungs and kidneys. Any structure or condition, temporary or permanent, that interferes with the blood flow also affects the homeostatic mechanisms that function to maintain the constancy of the internal environment.

The kidneys depend on an adequate blood supply for their normal functioning. Within the kidneys, microscopic structures filter the blood plasma. Hormonal control mechanisms regulate the reabsorption of filtered substances back into the blood stream. Thus, the kidneys regulate the content and volume of the blood, urine, and other fluids outside of the cells.

Sodium is one of the substances that is regulated by the kidney. The reabsorption of sodium, chloride, and water from the kidney tubules back into the blood and body fluids is regulated by sensitive control systems. When sodium is reabsorbed from the kidney tubules, water also passes back into the blood plasma. The water passing into the blood increases the blood volume, the output of the heart, and the blood pressure (Fig. 7-2).

The quantity of sodium in the blood and other extracellular fluids is regulated by a negative feedback system that involves the kidneys, the liver, and the adrenal gland. A compound called renin is produced by the kidney in response to decreased sodium or fluid content. Renin acts on liver angiotensinogen to produce angiotensin. When the outer part of the adrenal gland is acted upon by this angiotensin, it, in turn, produces a hormone called aldosterone. The net effect is to increase the reabsorption of sodium and water in order to maintain the heart rate and blood pressure. This typical negative feedback system is illustrated in Figure 7-3.

Any factor that influences the renin-angiotensin-aldosterone system may also affect the blood pressure. High blood pressure is one of the most serious complications of cardiovascular and kidney function. Hypertension, which is the sustained elevation of systemic ar-

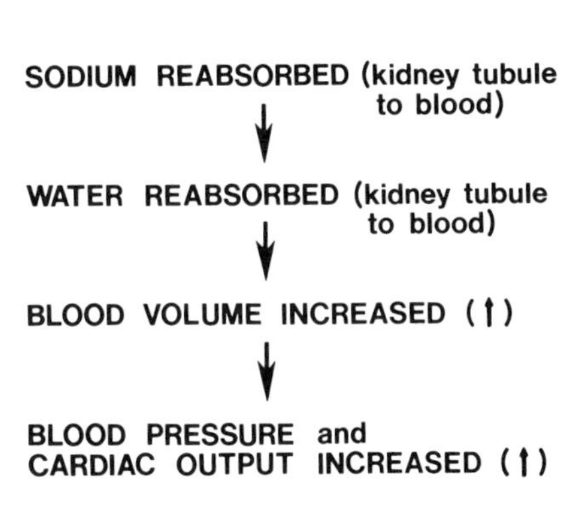

FIGURE 7-2 Relationship between sodium reabsorption and blood pressure.

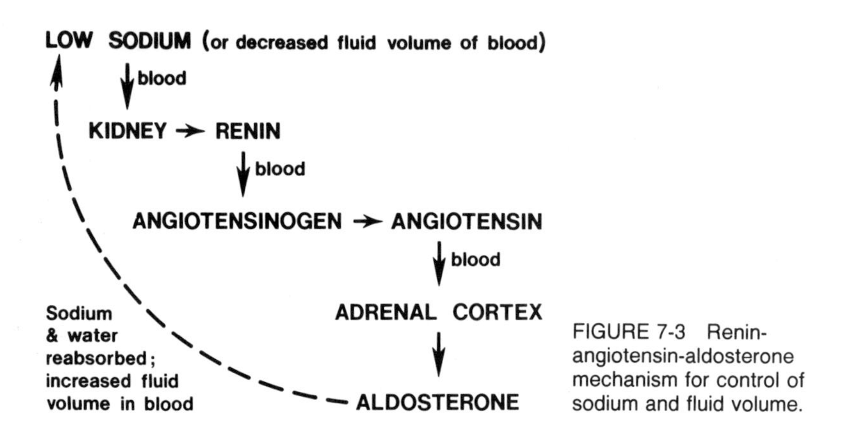

FIGURE 7-3 Renin-angiotensin-aldosterone mechanism for control of sodium and fluid volume.

terial pressure, can be a result of an interference with the aldosterone-mediated control of sodium reabsorption. Hypertension is also a predisposing condition for myocardial infarction (M.I.), a disease of the arteries that supply blood to the heart.

The causes of hypertension or high blood pressure include:

1. *Essential hypertension which is due to unknown causes. Smaller blood vessels are constricted, and this increases the blood pressure.*
2. *Diseases of the outer part of the adrenal gland.*
3. *Blood vessel constriction in the arteries that supply the kidney.*
4. *Diseases of the kidney.*
5. *Oral contraceptive steroids.*

When we take OCS, the amounts of estrogen and progesterone that circulate in our blood are increased. Some investigators (Mackay et al., 1973) think that estrogen and progesterone stimulate the production of renin and eventually increase the amount of aldosterone. Increased aldosterone promotes the retention of sodium and water, which may lead to hypertension.

Most women who use OCS do not become hypertensive. If women do become hypertensive while "on the Pill" most of them return to normal after discontinuing the medication. In most of us, the increases in renin concentration that may result from using OCS are suppressed by other mechanisms so that the rate of angiotensin and aldosterone production remains the same.

According to some estimates, hypertension symptoms may occur in as many as 15 to 18 percent of OCS users, although most estimates run closer to 5 percent. Those of us who use OCS should be convinced of the importance of frequent blood pressure checks, because what is true for the general population (Figure 7-4) applies more stringently to us.

FIGURE 7-4 Perform a death-defying act: Have your blood pressure checked.

Blood Clotting

Interference with blood flow in the form of a clot is an essential first step in the repair of our tissues after an injury. When our tissues are injured, several things happen. A substance called thromboplastin is released from the injured cells. At the same time damaged blood vessels deposit blood platelets, which gather at the site of the injury. Fibrin, released from the plasma by thromboplastin after a series of intermediate reactions, acts with the platelets to form a clot.

Clot formation in our injured blood vessels is obviously a homeostatic physiological response. It prevents severe blood loss. However, the formation of a clot within an intact, uninjured blood vessel is pathological, not protective. These pathological clot formations are called thromboses.

Studies that started in Great Britain (Vessey et al., 1976) during the 1960s showed that thromboses are a more common complication of OCS therapy than hypertension is. Over 50 percent of women who experienced cardiovascular complications related to the use of these medications reported thromboses. Less than 3 percent experienced pathological changes in blood pressure. We should realize that both problems, clotting and hypertension, are rare complications in women who use OCS. They are only seen in significant numbers when large populations of women are followed over significant periods of time. Such large-scale studies are difficult to carry out, but they are very valuable. They tell us not only that some women do develop cardiovascular problems after using OCS, but that these same women are more vulnerable if an additional event such as cardiovascular damage, surgical trauma, or inflammation occurs while they are on the pill.

Studies of women receiving estrogens as ERT therapy during the menopausal period of their lives have not turned up evidence of adverse effects similar to those in women using OCS. On the other hand, women who use large amounts of estrogen to suppress milk formation after childbirth face clotting dangers similar to those in women who are on OCS. Smoking cigarettes is an independent risk factor. Women who smoke and use OCS seem to have a greater tendency than OCS users who do not smoke (Pettiti et al., 1978). However, the physiological relationship between the risk factors has not been demonstrated.

The steroid hormones we use for contraception or for replacement therapy may affect important regulatory systems in our bodies. For some of us, the risks are greater because our metabolism is different. We need to know the risks that may be associated with the benefits we derive from effective contraception and from relieving some distressing symptoms of the menopause. To help women become familiar with these risks, the Food and Drug Administration has published the information reprinted in the following section.[1]

WHAT YOU SHOULD KNOW ABOUT ESTROGENS Estrogens are female hormones produced principally by the ovaries. The ovaries make several different kinds of estrogens. In addition, scientists have been able to make a variety of synthetic estrogens. As far as we know, all these estrogens have the same properties and therefore much the same usefulness, side effects, and risks. This [section] is intended to help you understand what estrogens are used for, the risks involved in their use, and how to use them as safely as possible.

[It] includes important information about estrogens in general, but not all the information. If you want to know more you can ask your doctor or pharmacist to let you read the professional package insert.

[1]Food and Drug Administration, "Information for the Patient about Estrogens," Washington, D.C.: Federal Register, August 1977.

Uses of Estrogen Estrogens are prescribed by doctors for a number of purposes, including:

1. *To provide estrogen during a period of adjustment when a woman's ovaries no longer produce it, in order to prevent certain uncomfortable symptoms of estrogen deficiency. (All women normally experience a decrease in the production of estrogens, generally between 44–55; this is called the menopause.)*
2. *To prevent symptoms of estrogen deficiency when a woman's ovaries have been removed surgically before the natural menopause.*
3. *To prevent pregnancy. (Some estrogens are given along with a progestagen, another female hormone; these combinations are called oral contraceptives or birth control pills. They will not be discussed in this leaflet.)*
4. *To treat certain cancers in women and men.*
5. *To prevent painful swelling of the breasts after pregnancy in women who choose not to nurse their babies.*

There is no proper use of estrogens in a pregnant women.

Estrogens in the Menopause In the natural course of their lives, all women eventually experience a decrease in estrogen production. This usually occurs between ages 45 and 55 but may occur earlier or later. Sometimes the ovaries may need to be removed before natural menopause by an operation, producing a "surgical menopause."

When the amount of estrogen in the blood begins to decrease, many women may develop typical symptoms: feelings of warmth in the face, neck, and chest or sudden intense episodes of heat and sweating throughout the body (called "hot flashes" or "hot flushes"). These symptoms are sometimes very uncomfortable. A few women eventually develop changes in the vagina (called "atrophic vaginitis") which cause discomfort, especially during and after intercourse.

Estrogens can be prescribed to treat these symptoms of the menopause. It is estimated that considerably more than half of all women undergoing the menopause have only mild symptoms or no symptoms at all and therefore do not need estrogens. Other women may need estrogens for a few months, while their bodies adjust to lower estrogen levels. Sometimes the need will be for periods longer than six months. In an attempt to avoid overstimulation of the uterus (womb), estrogens are usually given cyclically during each month of use, that is three weeks of pills followed by one week without pills.

Sometimes women experience nervous symptoms or depression during menopause. There is no evidence that estrogens are effective for such symptoms and they should not be used to treat them, although other treatment may be needed.

You may have heard that taking estrogens for long periods (years) after the menopause will keep your skin soft and supple and keep you feeling young. There is no evidence that this is so, however, and such long-term treatment carries additional risks.

ESTROGENS TO PREVENT SWELLING OF THE BREASTS AFTER PREGNANCY If you do not breast feed your baby after delivery, your breasts may fill up with milk and become painful and engorged. This usually begins about three to four days after delivery and may last for a few days to up to a week or more. Sometimes the discomfort is severe, but usually it is not and can be controlled by pain relieving drugs such as aspirin and by binding the breasts up tightly. Estrogens can sometimes be used successfully to try to prevent the breasts from filling up. While this treatment is sometimes successful, in many cases the breasts fill up to some degree in spite of treatment. The dose of estrogens needed to prevent pain and swelling of the breasts is much larger than the dose needed to treat symptoms of the menopause and this may increase your chances of developing blood clots in the legs or lungs or other parts of the body (see below, 4. Abnormal Blood Clotting). Therefore, it is important that you discuss the benefits and the risks of estrogen use with your doctor, before using estrogen, if you have decided not to breast feed your baby.

THE DANGERS OF ESTROGENS

1. Cancer of the Uterus. If estrogens are used in the postmenopausal period for more than a year, there is an increased risk of cancer of the endometrium (uterine lining). Women taking estrogens have roughly 5 to 15 times as great a chance of getting this cancer as women who take no estrogens. To put this another way, while a postmenopausal woman not taking estrogens has one chance in 1,000 each year of getting cancer of the uterus, a woman taking estrogens has 5 to 15 chances in 1,000 each year. For this reason it is important to take estrogens only when you really need them.

The risk of this cancer is greater the longer estrogens are used and also seems to be greater when larger doses are taken. For this reason it is important to take the lowest dose of estrogen that will control symptoms and to take it only as long as it is needed. If estrogens are needed for longer periods of time, your doctor will want to reevaluate your need for estrogens at least every six months.

Women using estrogens should report any irregular vaginal bleeding to their doctors; such bleeding may be of no importance, but it can be an early warning of cancer of the uterus. If you have undiagnosed vaginal bleeding, you should not use estrogens until a diagnosis is made and you are certain there is no cancer of the uterus. If you have had your uterus completely removed (total hysterectomy), there is no danger of developing cancer of the uterus.

2. Other Possible Cancers. Estrogens can cause development of other tumors in animals, such as tumors of the breast, cervix, vagina, or liver, when given for a long time. At present there is no satisfactory evidence that women using estrogens in the menopause have an increased risk of such tumors, but there is no way yet to be sure they do not; and one study raises the possibility that use of estrogens in the menopause may increase the risk of breast cancer many years later. This is a further reason to use estrogens only when clearly needed. While you are taking estrogens, it is important that you go to your doctor at least once a year for a physical examination. Also, if members of your family have had breast cancer or if you have breast nodules or abnormal mammograms (breast x-rays), your doctor may wish to carry out more frequent examinations.

3. Gall Bladder Disease. Women who use estrogens after menopause are two or three times more likely to develop gall bladder disease needing surgery than women who do not use estrogens. Birth control pills have a similar effect.

4. Abnormal Blood Clotting. Oral contraceptives increase the risk of blood clotting in various parts of the body. This can occur in different parts of the circulatory system causing thrombophlebitis (clot in the legs or pelvis), retinal thrombosis or optic neuritis (clots affecting vision, including blindness), mesenteric thrombosis (clots in the intestinal blood vessels), stroke (clot in the brain), heart attack (clot in a vessel of the heart) or a pulmonary embolism (clot which eventually lodges in the lungs). These can be fatal.

At this time use of estrogens in the menopause is not known to cause increased blood clotting. This has not been fully studied and there could still prove to be such a risk. It is recommended that if you have had clotting in the legs or lungs or a heart attack or stroke while you were using estrogens or birth control pills, you should not use estrogens (unless they are being used to treat cancer of the breast or prostate). If you have had a stroke or heart attack or if you have angina pectoris, estrogens should be used with great caution and only if clearly needed (for example, if you have severe symptoms of the menopause).

The larger doses of estrogen used to prevent swelling of the breasts after pregnancy have been reported to cause abnormal blood clotting as indicated above.

Special Warning about Pregnancy You should not receive estrogen if you are pregnant. If this should occur, there is a greater than usual chance that the developing child will be born with a birth defect, although the possibility remains fairly small. A female child may have an increased risk of developing cancer of the vagina or cervix later in life (in the teens or twenties). Every possible effort should be made to avoid exposure to estrogens during pregnancy. If exposure occurs, see your doctor.

OTHER EFFECTS OF ESTROGENS In addition to the serious known risks of estrogens described above, estrogens have the following side effects and potential risks which, if occurring, should be discussed promptly with your doctor.

1. *Nausea and vomiting. The most common side effect of estrogen therapy is nausea. Vomiting is less common.*
2. *Effects on breasts. Estrogens may cause breast tenderness or enlargement and may cause the breasts to secrete a liquid.*
3. *Effects on the uterus. Estrogens may cause benign fibroid tumors of the uterus to get larger. Some women will have menstrual bleeding when estrogens are stopped. But if the bleeding occurs on days you are still taking estrogens, you should report this to your doctor.*
4. *Effects on liver. Women taking oral contraceptives develop on rare occasions a tumor of the liver which can rupture, bleed into the abdomen, and cause death. So far, these tumors have not been reported in women using estrogens in the menopause, but you should report any swelling or unusual pain or tenderness in the abdomen to your doctor immediately.*
   Women with a past history of jaundice (yellowing of the skin and white parts of the eyes) may get jaundice again during estrogen use. If this occurs, stop taking estrogens and see your doctor.
5. *Other effects. Estrogens may cause excess fluid to be retained in the body. This may make some conditions worse, such as epilepsy, migraine, heart disease, or kidney disease. Mental depression or high blood pressure may occur. A spotty darkening of the skin, particularly of the face, is possible and may persist.*

SUMMARY Estrogens have important uses, and they may have serious risks as well. You must decide, with your doctor, whether the risks are acceptable to you in view of the benefits of treatment. Except where your doctor has prescribed estrogens for use in special cases of cancer of the breast or prostate, you should not use estrogens if you have cancer of the breast or uterus, are pregnant, have undiagnosed abnormal vaginal bleeding, clotting in the legs or lungs or have had a stroke, heart attack or angina, or clotting in the legs or lungs in the past while you were taking estrogens.

You can use estrogens as safely as possible by understanding that your doctor will require regular physical examinations while you are taking them and will try to discontinue the drug as soon as possible and use the smallest dose possible. Be alert for signs of trouble including:

1. *Abnormal bleeding from the vagina.*
2. *Pains in the calves or chest or sudden shortness of breath, or coughing blood (indicating possible clots in the legs, heart, or lungs).*
3. *Severe headache, dizziness, faintness, or changes in vision (indicating possible developing clots in the brain or eye).*
4. *Breast lumps (you should ask your doctor how to examine your own breasts).*
5. *Jaundice (yellowing of the skin).*
6. *Mental depression.*

Based on his or her assessment of your medical needs, your doctor has prescribed this drug for you. Do not give the drug to anyone else.

HORMONES AND BEHAVIOR

The behavior of women, influenced as we are by periodic fluctuations in our hormone levels, has been commented upon since early antiquity. The Greeks and Romans referred to the "wandering uterus" as a cause for strange behavior in menstruating women. During the Middle Ages, witchcraft was invoked to explain unusual behavior in women afflicted with "the curse" of menstrual bleeding. In some primitive societies women are still isolated from the rest of the tribe during the period of menstrual bleeding.

Today and here we may hear ourselves or other women making statements like these:

"I felt great *during all the time I was pregnant."*

"After the baby was born, there was a time when I cried at any little thing."

"Taking the pills (OCS) made me feel so depressed that I had to stop taking them."

"Just before my period I become irritable, depressed and just plain bitchy."

We may have heard or made statements like these. Yet, we wonder how and why the ovarian hormones, estrogen and progesterone, may affect our feelings and behavior. Perhaps we have observed that our friends are different as the hormonal influences on their bodies

FIGURE 7-5 Women take their children to the doctor more frequently during the premenstrual and menstrual phases of the menstrual cycle.

change with time. They may have observed the same thing about us. Sometimes we take these mood and behavioral changes lightly and with humor. Sometimes the things we feel and do are not so funny. The cartoons in Figures 7-5, 7-6, and 7-7 refer to some behavioral aspects of the hormonal changes in our bodies.

Misconceptions about Behavior

Many of our ideas about behavior may be based on wrong assumptions which include:

1. *The idea that a one-to-one relationship exists between cause and effect. Even animal experiments do not always warrant this assumption, and so we should not believe that a high or low level of a certain hormone is the* sole *cause of an observed behavior.*

FIGURE 7-6 "It must be *that* time of the month again!"

FIGURE 7-7 "I wonder what she's like during *that* time of the month?"

2. *Equating the behavior itself with the meaning of the behavior. We know, for example, that children and adults sometimes display uncontrolled anger. This is often called a temper tantrum. The behavior may not always signify anger; it may mean that someone craves attention.*
3. *Oversimplification. We can attribute behavior to broad general terms like "aggression," "anxiety," "depression," "learning," and so forth. But each one of these terms is not a single thing. Each consists of multiple meanings with many inputs, both factual and emotional. These terms constitute wholes that cannot be fully understood merely by an enumeration of the parts.*

If we try to understand our behavior, as it may be influenced by the hormones in our bodies, we need to beware of these false assumptions. Behavior is a complex product of many interacting factors. Studies of the behavior and moods of women during different hormonal phases of our lives often fail to take that complexity into account. We ourselves know that we often *do* (behavior) what we don't *feel like doing* (mood) simply because we know that we have to do it. On the other hand, we cannot always do what we feel like doing because of the restrictions of time and our obligations to other people.

Evaluating Behavioral Findings

A number of research studies have attempted to assess the effects of our hormone cycles on our behavior patterns. Sexual behavior, propensity to suicide, motivation, and behavioral changes during menstruation are among the dimensions that have been researched. Because the results of such studies make fascinating reading, they are frequently reported in some of our favorite magazines. Before we agree with the authors' conclusions, however, we should ask ourselves some basic questions.

First, how were the women assigned to different ovulatory cycle phases? All the methods that were described in Chapter 6 for determining the time of ovulation could be used for this purpose, including BBT records, the calendar, vaginal cytology, cervical mucus changes, and the measurement of hormones from the pituitary (LH and FSH) and the ovary (estrogen and progesterone). There is considerable woman-to-woman variation in the *amount* of each hormone secreted during each phase of the cycle. This should be taken into account in the summarizing conclusions; for example, one woman's mid-cycle estrogen peak may measure 100 units while another's could be considerably more or less. These individual variations in hormone levels may cause different effects, even though each woman is at the same phase of her cycle.

Other important questions to ask include, Does this study avoid assumptions that assign an oversimplified cause-and-effect relationship? Does it equate the behavior with the meaning of the behavior? Are complex concepts such as "depression," "anxiety," and "human sexual behavior" oversimplified?

Sexual Behavior Human sexual behavior during different times of the ovulatory cycle was a subject of interest long before Masters and Johnson took sex into the laboratory. Some early studies used imprecise methods for determining menstrual cycle stages. Often they relied solely upon the ability of a woman to remember her sexual response in relationship to the stage of her monthly cycle. Most of us, however, are more apt to remember our response in relationship to the setting and the behavior of our partner. At the present time, no clear picture emerges from studies on the relationship between our sexual behavior as it relates to the stage of the ovulatory cycle (Udry and Morris, 1977).

Suicidal Behavior The question of whether women are more prone to take their own lives during certain phases of the ovulatory cycle continues to attract research interest. Perhaps the intention to commit suicide is a measurement of the depression that may affect some of us during some phases of the cycle, usually just before menstrual bleeding. Conclusions about the relationship between cycle phase and suicide attempts do not all agree, and this should not surprise us. A woman's decision to end her own life is a deeply personal one. It may be arrived at only after many complex and interrelated aspects of her life are weighed and considered. Factors entirely unrelated to the ovulatory cycle may be involved. For example, the depression that precedes such a decision may be one of the effects of excessive alcohol consumption (see Chapter 10). Traumatic life events such as the death of a loved one, the infidelity of a spouse or lover, or the irresponsible selfishness of a son or daughter may also lead to suicidal depression. The hormonal influences on a woman's decision to commit suicide can only be weighed and measured correctly when these and other nonhormonal influences have been accounted for.

Motivation Perhaps in all the studies of the effects of hormones on behavior, too little emphasis has been placed on motivation. This fact was underscored in a study of the effects of the hormone changes during the ovulatory cycle on the academic performance of college women (Bernstein, 1977). These women, academically advanced and apparently highly motivated, showed no drop in their intellectual performance, as measured by test scores, that could be related to their premenstrual or menstrual periods. This finding is important to us. We can be encouraged to know that we can do what we want to do, regardless of the hormonal influences on our bodies.

Menstrual Distress One of the best-known measurements of the unpleasant symptoms that may occur during the ovulatory cycle is the Menstrual Distress Questionnaire designed by Moos (1968, 1969). A list of 47 symptoms in this questionnaire, which appears in Table 7-4, was obtained by asking women about their symptoms and by referring to the literature on the subject. Table 7-5 summarizes the responses of over 800 young married women to this questionnaire. There are certain symptoms many women experience during their ovulatory cycles; some are more frequent for some of us than for others. Women who are on OCS have more regular cycles (Fig. 7-8, page 162) and seem to experience fewer debilitating symptoms associated with menstruation (Fig. 7-9, page 162). While the OCS relieve a number of symptoms associated with menstrual distress,

TABLE 7–4 Menstrual Distress Questionnaire (MDQ) Symptom Scales

1. *Pain*
   - Muscle stiffness
   - Headache
   - Cramps
   - Backache
   - Fatigue
   - General aches and pains
2. *Concentration*
   - Insomnia
   - Forgetfulness
   - Confusion
   - Lowered judgment
   - Difficulty concentrating
   - Distractible
   - Accidents
   - Lowered motor coordination
3. *Behavioral change*
   - Lowered school or work performance
   - Take naps; stay in bed
   - Stay at home
   - Avoid social activities
   - Decreased efficiency
4. *Autonomic reactions*
   - Dizziness, faintness
   - Cold sweats
   - Nausea, vomiting
   - Hot flashes
5. *Water retention*
   - Weight gain
   - Skin disorders
   - Painful breasts
   - Swelling
6. *Negative affect*
   - Crying
   - Loneliness
   - Anxiety
   - Restlessness
   - Irritability
   - Mood swings
   - Depression
   - Tension
7. *Arousal*
   - Affectionate
   - Orderliness
   - Excitement
   - Feelings of well-being
   - Bursts of energy, activity
8. *Control*
   - Feeling of suffocation
   - Chest pains
   - Ringing in the ears
   - Heart pounding
   - Numbness, tingling
   - Blind spots, fuzzy vision

Source: Rudolph H. Moos, "Assessment of Psychological Concomitants of Oral Contraceptives," in *Metabolic Effects of Gonadal Hormones and Contraceptive Steroids,* ed. H.A. Salhanick, D.M. Kipnis, and R. L.Vande Wiele (New York: Plenum Press, 1969), p. 678, by permission of Plenum Publishing Corporation.

the extent to which the symptoms are ameliorated varies from subject to subject and from study to study. The psychological ramifications of OCS use and those resulting from menstrual distress may be separate problems to be investigated in future studies.

FUTURE STUDIES Studies yet to be done on the effects of hormones on the behavior of women may include a variety of subjects who have different life-styles (single, divorced, living with a partner, with children, or childless). These women may also be in different stages of their lives as far as their hormone levels are concerned (ovulatory, pregnant, menopausal, with and without ERT). Hormone levels can be measured accurately. Behavior is more difficult to measure because, like the human beings in whom it is studied, it is so complex. We know, in a general way, that hormonal changes may *influence* our behavior. We also know, from having lived with these hormonal influences that we ourselves determine what we *do* in spite of, or because of what we *feel.*

TABLE 7–5 Percent of Women with Mild, Moderate, Strong, and Severe Complaints on Selected Symptoms on MDQ

	Menstrual		*Premenstrual*	
Scale Sympton	*Mild Moderate*	*Strong Severe*	*Mild Moderate*	*Strong Severe*
Pain				
Cramps	35.6	11.0	12.3	1.9
Backache	31.9	8.1	20.7	3.7
Concentration				
Difficulty concentrating	12.3	1.9	12.9	1.2
Accidents	11.7	1.3	12.6	1.8
Behavior Change				
Take naps; stay in bed	22.4	3.8	10.4	2.1
Decreased efficiency	23.4	2.6	16.0	2.6
Autonomic Reaction				
Dizziness, faintness	9.7	1.5	3.8	0.9
Nausea, vomiting	5.3	1.2	3.6	0.7
Water Retention				
Weight gain	20.7	2.2	30.8	3.1
Swelling	31.4	4.0	30.4	5.1
Negative Affect				
Irritability	40.5	8.4	39.2	13.0
Depression	27.5	7.1	33.4	9.5
Arousal				
Well being	23.6	4.4	20.5	5.0
Energy, activity	19.4	4.5	20.0	6.6
Control				
Suffocation	1.1	0.5	1.2	0.1
Ringing in ears	1.8	0.7	1.6	0.4

Source: Rudolph H. Moos, "Assessment of Psychological Concomitants of Oral Contraceptives," in *Metabolic Effects of Gonadal Hormones and Contraceptive Steroids,* ed. H.A. Salhanick, D.M. Kipnis, and R. L.Vande Wiele (New York: Plenum Press, 1969), p. 679, by permission of Plenum Publishing Corporation.

PARTICIPATING IN RESEARCH ON HORMONES, METABOLISM, AND BEHAVIOR

Some questions that came up during the studies (Moos, 1969) of menstrual distress in women using OCS compared to those who did not use them included:

1. *Do the OCS exert the same influences on the emotions as naturally produced estrogen and progesterone?*
2. *Do the OCS interact with other medication that women may be taking?*
3. *Are there genetically determined differences in the pathways utilized for metabolizing steroid medications?*

FIGURE 7-8 "Our sales are a lot more predictable now that they're all on the Pill."

FIGURE 7-9 Effect of oral contraceptive steroids on work performance and efficiency. Percent of women reporting moderate, strong, or acute symptoms in oral contraceptive and no oral contraceptive groups in menstrual (M), premenstrual (PM) and intermenstrual (IM) phases of most recent menstrual cycle. From Rudolf H. Moos, "Assessment of Psychological Concomitants of Oral Contraceptives," in *Metabolic Effects of Gonadal Hormones and Contraceptive Steroids,* ed. Hilton A. Salhanick, David M. Kipnis, and Raymond L. Vande Wiele (New York: Plenum Press, 1969), p. 688, by permission of Plenum Publishing Corporation.

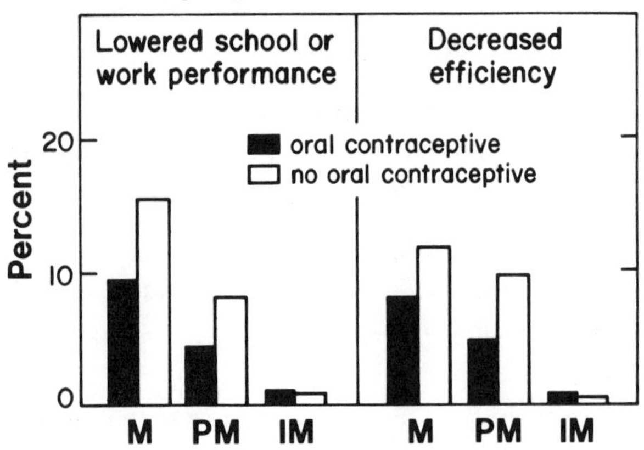

We know that further studies are necessary to answer these and other questions and to increase our knowledge about what hormones do to our bodies. Some of us may be asked to participate in such studies. If we are poor, or if we are hospital patients in wards or clinics, we may be more likely to be subjects of research studies. Because such studies often involve health risks as well as potential benefits, risk-benefit relationships are determined by the persons who design the research study. Since the OCS and ERT have been used for many years, their risk-benefit relationships are fairly well known. Those of us who use these medications have made a judgment that, for us, the use of the OCS or ERT in terms of the benefits we derive outweighs the potential risks. If tests are to be carried out on us we should likewise seek to understand and evaluate for ourselves the risk-benefit relationships of any medications or procedures involved. We should also be familiar with the reputation of the person who is directing the research. If we have confidence in the researcher then we also believe that what he or she is asking us to do will help us and other women like ourselves, and that, as a result of this study, valuable information on the metabolic and/or behavioral effects of gonadal hormones will be gained.

It is our right to understand what we are volunteering for. Unfortunately, this right is not always respected.

At a respected university medical center, 51 women were interviewed who had been the subjects in a study of the effects of a new labor-inducing drug. Although the women had signed a consent form, often in the hectic course of the admitting procedure or in the labor room itself, 20 of them (39%) learned only from the interview that they were the subjects of research. The interview was held after the drug infusion had been started or after the delivery.

Among the women who did know that they were subjects for research, most of them did not understand that there might be dangers related to the research. They did not understand that they were also going to be watched and subjected to special test procedures unrelated to the delivery of the baby. In addition to this, they did not understand that they were free not *to participate; four of the women said that they would have refused to participate if they had known there was any choice. Many of the women had been referred for the studies by their private physicians, but instead of being informed that an experimental drug was to be administered, they were told it would be a "new" drug. They trusted their doctor and assumed that "new" meant "better." [Barber, 1976]*

Those of us who may be asked to participate in a study of effects of OCS or ERT have already chosen to take the medication (or not to take it if we are asked to be control subjects). However, we cannot freely choose to participate in the research unless we understand what the research is all about. Because researchers are required by law to obtain *informed consent* from each participant in a research project, we will be asked to sign a consent form. This form should include:

1. *the purpose of the research*
2. *the procedures which are involved in it*
3. *the risks which might be dangerous to our health*
4. *the benefits to us*
5. *a statement that we are free to withdraw from the research project whenever we so desire*
6. *the invitation to ask as many questions as we want to ask.*

Provision 6 seems to be an excellent safeguard; however, it may be of little or no help to us if we do not understand the document in the first place and if we are too intimidated to admit it.

Gray and his associates (1978) analyzed the reading level of consent forms that had been approved by research boards of scientific institutions. This analysis showed that the reading level of the consent form was very difficult (scientific/professional) in 20 percent of the consent forms. A difficult (scholarly/academic) level was found in over 50 percent of the others. A fairly difficult reading level (above that in *Time* magazine) was characteristic of 10 to 20 percent of the remaining consent forms. Thus, 80 to 90 percent of the consent forms were written on a level so difficult that most persons who were asked to sign them could not understand what they were signing. Most of us find it very difficult to ask a question about something we cannot begin to understand, particularly if we are in some kind of awe of the person in the white coat. Uneducated women are even more vulnerable if they are asked to be subjects for research.

Communicating clearly to us about what we are volunteering for is the obligation of the research scientist. If the scientist's research and medical training has not included effective communication skills, someone who has these skills should be assigned to interview the subjects. The ethics of research on human subjects recognizes our right to understanding and voluntary participation.

We may one day be asked to participate in biomedical-clinical studies. It is our right to understand the purpose of the research and the procedures that will be followed. We can hope that research persons who ask better questions may give us better answers to problems which concern our hormones, metabolism and behavior.

REFERENCES

PHYSIOLOGICAL ACTION OF STEROID HORMONES

Briggs, Michael H. "Human Metabolism and Steroid Contraception," *Australian Family Physician,* Special Issue, March 1977, pp. 23–28.

Coope, Jean; Jean M. Thompson; and L. Poller. "Effects of 'Natural Oestrogen' Replacement Therapy on Menopausal Symptoms and Blood Clotting," *Br. Med. J.* 4 (October 18, 1975): 139–43.

Cox, B. D., and D. P. Calhane. "Changes in Plasma Amino Acid Levels during the Human Menstrual Cycle and Early Pregnancy," *Hormones and Metabolic Research* 10 (1978): 428–33.

Cudworth, A. G. "Carbohydrate Metabolism in the Menstrual Cycle," *Br. J. Obstet. Gynaec.* 82 (1975): 162–69.

Ferguson, Ann. "Oral Contraceptives, Thromboembolic Disease, and Hypertension: A Review," *Journal of the Medical Association of the State of Alabama,* September 1977, pp. 49–55.

Fuch, R. H. "The Relationship between a Mother's Menstrual Status and Her Response to Illness in Her Child," *Psychomatic Medicine* 37, no. 5 (1975): 388–94.

Hilliard, George D., and Henry J. Morris. "Pathologic Effects of Oral Contraceptives," *Recent Results in Cancer Research* 66 (1979): 49–71.

Mackay, E. V.; S. K. Khoo; and N. A. Shah. "Reproductive Steroids and the Circulatory System with Particular Reference to Hypertension," *Obstetrical and Gynecological Survey* 28, no. 3 (1973): 49–55.

Miale, J. B., and J. W. Kent. "The Effects of Oral Contraceptives on the Results of Laboratory Tests," *Am. J. Obstet. Gynecol.* 120, no. 2 (1974): 264–72.

Morrison, John A.; Kathe Kelly; Margo Mellies; Ido de Groot; Philip Khoury; Peter S. Gartside; and Charles J. Glueck. "Cigarette Smoking, Alcohol Intake, and Oral Contraceptives: Relationships to Lipids and Lipoproteins in Adolescent School Children," *Metabolism* 28, no. 11 (November 1979): 1166–70.

Petitti, Diana B.; John Wingard; Frederick Pellegrin; and Savitri Ramcharan. "Oral Contraceptives, Smoking, and Other Factors in Relation to Risk of Venous Thromboembolic Disease," *American Journal of Epidemiology* 108, no. 6 (1978): 480–84.

Rose, D. P., and P. W. Adams. "Oral Contraceptives and Tryptophan Metabolism: Effects of Oestrogen in Low Dose Combined with a Progestogen and of a Low-Dose Progestogen (Megastrol Acetate) Given Alone," *Journal of Clinical Pathology* 25 (1972): 252–58.

Salhanick, Hilton A.; David M. Kipnis; and Raymond L. Vande Wiele. *Metabolic Effects of Gonadal Hormones and Contraceptive Steroids.* New York: Plenum Press, 1969.

Southam, Anna L., and Florante P. Gonzaga. "Systemic Changes during the Menstrual Cycle," *Am. J. Obstet. Gynecol.* 91, no. 1 (January 1, 1965): 142–65.

Spellacy, William N.; William C. Buhi; and Sharon A. Birk. "Three-Year Prospective Study of Carbohydrate Metabolism in Women Using Ovulen," *Southern Medical Journal* 70, no. 10 (October 1977): 1188–90.

———. "Effects of Norethindrone on Carbohydrate and Lipid Metabolism," *Obstet. Gynecol.* 46, no. 5 (November 1975): 560–63.

Udry, J. R., and Naomi M. Morris. "The Distribution of Events in the Normal Menstrual Cycle," *Journal of Reproduction and Fertility* 51 (1977): 419–25.

Vander, Arthur J.; James H. Sherman; and Dorothy S. Luciano. *Human Physiology: The Mechanisms of Body Function,* 2nd ed. New York: McGraw-Hill Book Company, 1975.

Vessey, Martin; Sir Richard Doll; Pichard Peto; Bridget Johnson; and Peter Wiggins. "A Long-Term Follow-Up Study of Women Using Different Methods of Contraception—An Interim Report," *Journal of Biosocial Science* 8 (1976): 375–427.

von Kaulla, E.; W. Droegemueller; and K. N. von Kaulla. "Conjugated Estrogens and Hypercoagulability," *Am. J. Obstet. Gynecol.* 122, no. 6 (July 15, 1975): 688–92.

Wallace, Robert B.; Joanne Hoover; Dale Sandler; Basil M. Rifkind; and Herman A. Tyroler. "Altered Plasma Lipids Associated with Oral Contraceptive or Oestrogen Consumption," *Lancet,* July 2, 1977, pp. 11–14.

HORMONES AND BEHAVIOR

Bernstein, Barbara Elaine. "Effect of Menstruation on Academic Performance Among College Women," *Archives of Sexual Behavior* 6, no. 4, (1977): 289–96.

McCauley, Elizabeth, and Anke A. Ehrhardt. "Female Sexual Response: Hormonal and Behavioral Interactions," *Primary Care* 3, no. 3 (September 1976): 455–76.

McKearney, James W. "Asking Questions about Behavior," *Perspectives in Biology and Medicine* 20 (Autumn 1977): 109–19.

Moos, Rudolph H. "Psychological Aspects of Oral Contraceptives," *Archives of General Psychiatry* 19 (July 1968): 87–94.

———. "Assessment of Psychological Concomitants of Oral Contraceptives," in *Metabolic Effects of Gonadal Hormones and Contraceptive Steroids,* ed. H. A. Salhanick, D. M. Kipnis, and R. Vande Wiele. New York: Plenum Press, 1969.

Pallis, D. J., and T. A. Holding. "The Menstrual Cycle and Suicidal Intent," *Journal of Biosocial Science* 8 (1976): 27–33.

PARTICIPATING IN RESEARCH

Barber, Bernard. "The Ethics of Experimentation with Human Subjects," *Scientific American* 234, no. 2 (February 1976): 25–31.

Gray, Bradford H.; Robert A. Cooke; and Arnold S. Tannenbaum. "Research Involving Human Subjects," *Science* 201 (September 22, 1978): 1094–101.

8
Cancer in Women

What happens when a woman first learns she has cancer? The following excerpt from Marvella Bayh's account of her decade-long battle with cancer portrays a typical reaction: disbelief.

I wrote in my diary: Scared. Pray breast condition not cancerous.

It was.

I could not believe it. With only a 10 percent chance of it being so, with no palpable lump, no indications from the x-rays, I had breast cancer. I was *that statistic: 1 out of every 13 American women.*[1]

The expressed feelings of one woman could become our own at some time in our lives. The American Cancer Society tells us that one out of every five women in this country will have cancer of some kind at one time during her life. Cancer, the leading cause of death in women from age 30 to 54, may affect many different organs in our bodies. The relative frequency of the main forms of cancer in women is summarized in Figure 8-1.

Most of us, if we are honest with ourselves, will admit that we are frightened by the idea of cancer, particularly if this abstract "thing" should ever become a personal experience. Knowing more about it may at least give us a focus for our fears, and may enable us to put some of these fears aside. Knowing what to expect from cancer is one way we can maintain control of our lives and health.

[1]Marvella Bayh and Mary Lynn Kotz, *Marvella: A Personal Journey* (New York: Harcourt Brace Jovanovich, 1979). Reprinted by permission of Harcourt Brace Jovanovich and The Sterling Lord Agency, Inc. © 1979.

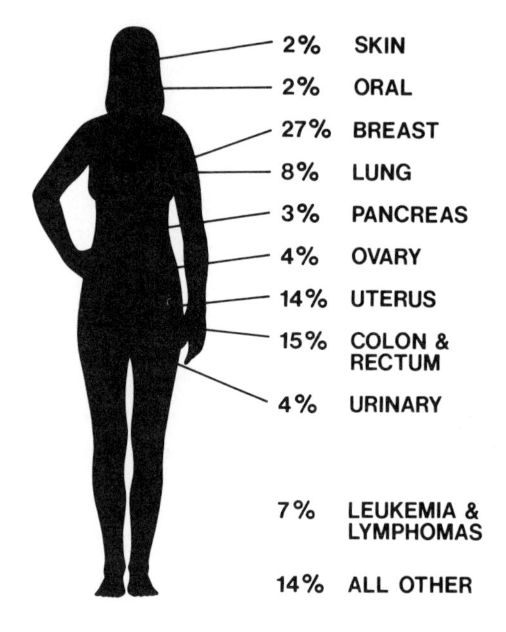

FIGURE 8-1 Cancer in women. Percentages indicate the relative frequency of each form of cancer in women. Of every 100 cancers, 2 are skin cancers, 2 are oral cancers, 27 are breast cancers, and so on. From "Cancer Facts and Figures" (New York: American Cancer Society, Inc., 1980), p. 2. Courtesy of the American Cancer Society.

WHAT IS CANCER?

Although cancer can occur in many different parts of our bodies, it has the same basic characteristic wherever it occurs: the unrestrained and independent proliferation of cells. This uncontrolled production of new cancer cells may take place at the expense of normal cells. Cancer, wherever it may appear in the organs and tissues of our bodies—in our breast cells, our uterine cells, or our skin cells—begins as a disease of cells.

Cancer as a Disease of Cells

Our bodies are made up of cells. These cells, which perform diverse metabolic functions, have different structures adapted to these functions. Our lives began with cells. These specialized reproductive cells, ovum and sperm, are called germ cells or gametes. The zygote formed by the union of these gametes contained the chromosomes contributed by both our parents. The cells of our bodies were produced by divisions of the zygote and subsequent specialization (differentiation) of our embryonic cells. Cell division and cell specialization are still going on in our bodies. What regulates these processes?

In order to understand something about cancer, we need to understand cell division and cell differentiation, because the cancer cell is a normal cell that has somehow escaped the control mechanisms regulating these processes. Cancer cells reproduce more rapidly than normal cells. They have been permanently altered so that

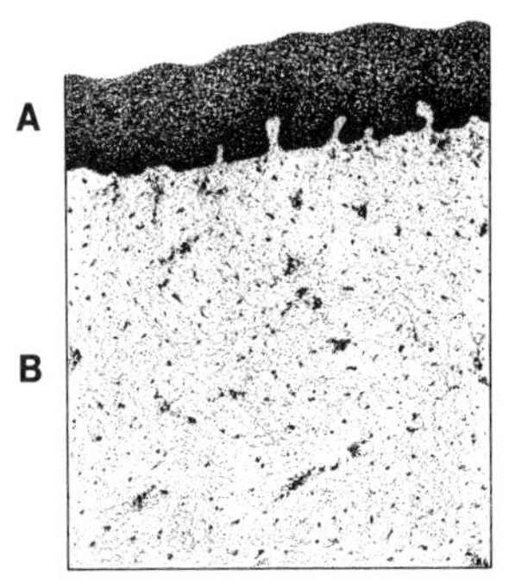

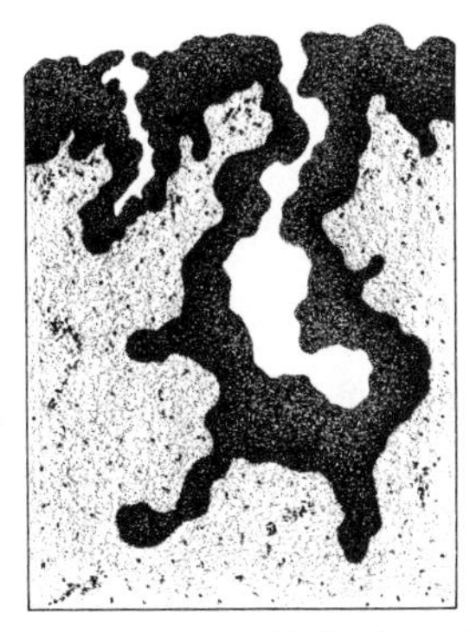

FIGURE 8-2 The endometrium before and after multiplication of cancer cells. Cells from the upper layer of the uterus (A) become malignant and grow down into the lower layer (B). The structure and function of both layers is disrupted.

our bodies have no adequate defense mechanisms against them. The cells of malignant tumors invade and destroy normal tissue; for example, observe in Figure 8-2 the difference between normal uterine tissue (1) and uterine tissue that has been invaded by cancer cells (2). In these histologically processed tissues, the cells in layer A have grown out of control down into layer B, so that the structure (and function) of the uterine lining is completely changed. Cancers can also metastasize; that is, the cancer cells can move through the circulatory system to new sites in our bodies where new tumors may develop.

Cancer cells continue to multiply and grow at the expense of the tissue they have invaded. They grow and divide independently without contributing anything to the functioning of our bodies; in this sense they are autonomous. Sickness and death in the individual are possible results of cancer's unrestrained growth. Unlike normal cells, which function as parts of an integrated organ system, cancer cells seem to "march to a different drummer." They appear to have different instructions from those of the normal cells from which they are derived. How did cancer cells get their new and destructive information? If we could answer that question with any degree of precision, we could say that we understood cancer, and were able to control it. We don't and we can't—so far.

Some Approaches to Understanding Cancer

The Cell Cycle Since cancer is a cell disease, characterized by unrestrained growth, we might understand it better if we examine the processes by which our cells reproduce themselves. The cell cycle (Fig. 8-3) is a series of events in cell reproduction divided into several different phases.

THE NORMAL CELL CYCLE

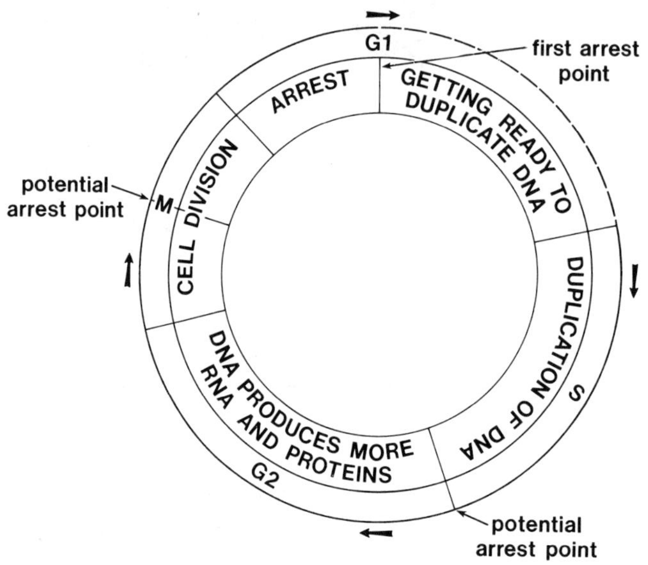

FIGURE 8-3

1. *G1 phase. Cells produce material necessary to duplicate DNA.*
2. *S phase. DNA is duplicated in preparation for cell division.*
3. *G2 phase. Substances required for nuclear division are produced.*
4. *M phase. Mitosis or cell division occurs in four stages: prophase, metaphase, anaphase, telophase.*

During the normal cell cycle several blocks to further division occur. These blocks are removed in tumor cells, however, allowing their unrestrained growth. For example, during the normal cell cycle, the chromosomes are duplicated only once; however, when the control mechanisms are removed, several sets of chromosomes may be produced (Fig. 8-4). Understanding the nature of cancer requires a knowledge of the *control mechanisms* that regulate normal cell growth.

DNA and Protein Synthesis The function of chromosomal DNA in the production, through RNA, of specific proteins was summarized in Chapter 2. There is a possibility, however, of wrong information being incorporated into the genetic instructions. This, in turn, could lead to the synthesis of proteins that do not contribute adequately to the function or structure of the cells, tissues, and organs in which they are located. The paired bases of DNA are duplicated during DNA synthesis (cell cycle, S phase). If chemical changes occur in these bases, they will not function correctly during DNA duplication. Similarly, changes in DNA can bring about changes in RNA and, thus, in the protein synthesis which DNA directs.

Some agents that are implicated in cancer, such as x-rays, ultraviolet irradiation, and certain chemicals, may actually alter the structure of DNA and ultimately of the cell proteins. Viruses can incorporate themselves into the DNA of the host organism, and thus change the developmental directions contained in the genetic material. Some changes in the protein produced could affect the cell membrane. Cancer cells do not adhere to other cells, nor are they prevented from growing by contact with other cells (contact inhibition). Some of the proteins produced by the cell are antibodies that operate in the immune response. It becomes evident that understanding the nature of cancer will involve following the way genetic information is translated into protein synthesis.

CELL DIFFERENTIATION All of our chromosomes, with their genes, are present in all our cells, including tumor cells. Why then are our cells different from each other? How have they become specialized to perform so many diverse functions in our bodies—the secretion of hormones, the conduction of nerve impulses, the contraction of muscles, and many others? How do some cells become tumor cells? Geneticists and developmental biologists believe that our development is a matter of selective gene derepression. Another way of saying this is that the genes do not produce the same proteins in all our cells because some gene activity is held in check by control mechanisms within the body. Genes can only function when these controlling factors are removed, as they are, sequentially, in normal cells. Some factor, not clearly identifiable at the present time, but present in the environment of the genes, removes the repressing factors from some of the genes and they begin to synthesize the proteins characteristic of a particular cell type (Fig. 8-5). Apparently, the genes that characterize the production of a cancer are present, but inactive, in our chromosomes. Some external environmental influence may trigger their production of abnormal proteins. It is pos-

FIGURE 8-4 Chromosomes and cancer.

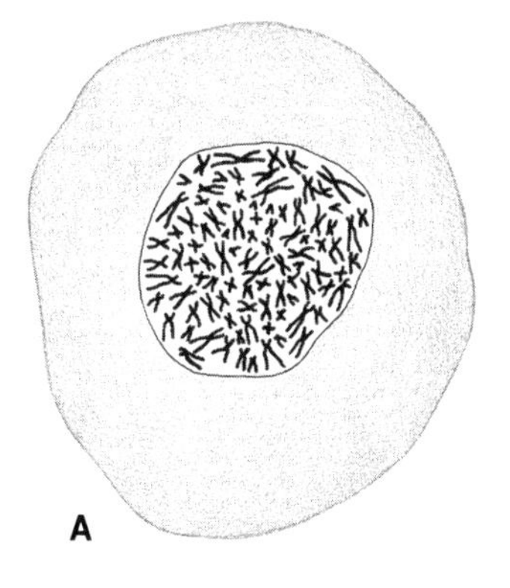

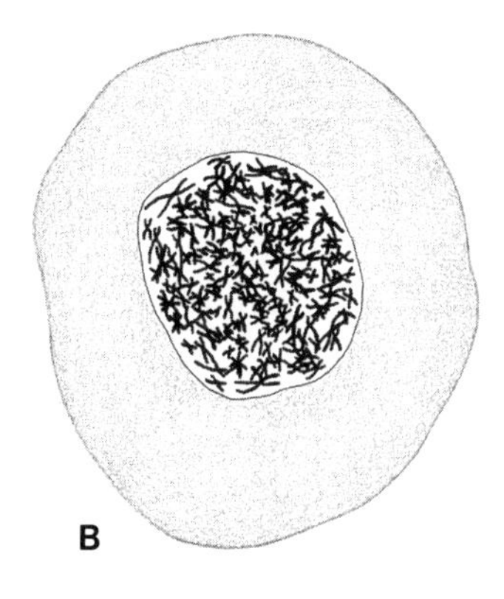

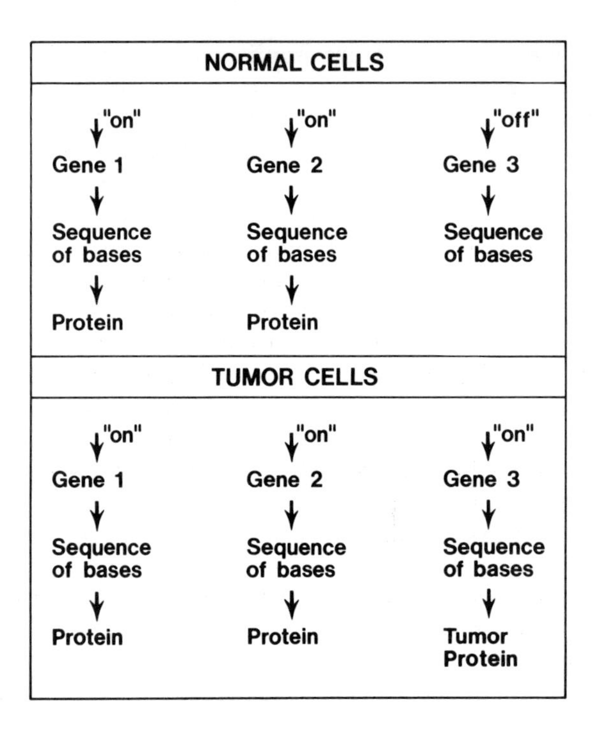

FIGURE 8-5 Gene derepression and protein synthesis in normal cells and cancer cells.

sible that the tumorous state results from the persistent activation of biosynthetic systems that are repressed in normal tissue.

The genes that are activated when normal cells become cancer cells do not affect the other genes. It has been shown in frogs that the nuclei of cancer cells still have all the chromosomal information necessary to direct the development of the complete animal. As a matter of fact, if the nucleus from a frog tumor cell is transplanted into a frog egg from which the normal nucleus has been removed, a fully normal frog tadpole can develop (Fig. 8-6).

THE IMMUNE RESPONSE Our bodies have a unique and sensitive system, the immune response, for recognizing foreign substances, whether these substances are viruses, bacteria, or tissue transplants. The immune system is also involved in the destruction of abnormal cells such as cancer cells. Somehow our bodies fight these invaders because they are "non-self." How does this happen?

Immune mechanisms in our bodies are of two types, non-specific and specific, both directed by white blood cells (Fig. 8-7). Specific immune responses are homeostatic mechanisms in which the stimulus is the foreign cell or a subcellular part of it, and the effectors are specialized white blood cells, the lymphocytes. These lymphocytes are of two types, the T-cells which attack foreign materials directly and the B-cells which produce antibodies to deal with them. Under

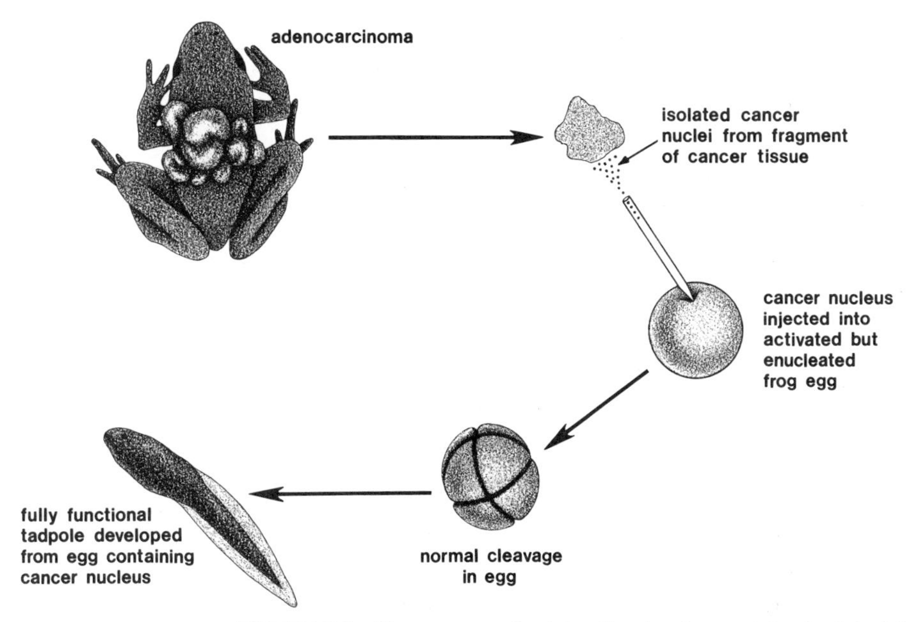

FIGURE 8-6 The cancer cell retains the developmental potential of the normal cell. Adapted from *The Story of Cancer,* 1977, written by Armin C. Braun, with permission of publishers, Addison-Wesley, Advanced Book Program, Reading, Massachusetts, U.S.A.

normal circumstances anything which is non-self in our bodies is eliminated by the activities of the immune system. Somehow, however, the cancer cells fool the immune system. Or it may happen that the cancer cells become established because the immune system isn't operating properly. In this case the failure of the immune system contributes to the success of the cancer. In order to use the immune system to fight cancer, scientists need to know more about the diverse responses of the immune system in health and disease.

We don't know yet what causes cancer and why some of us

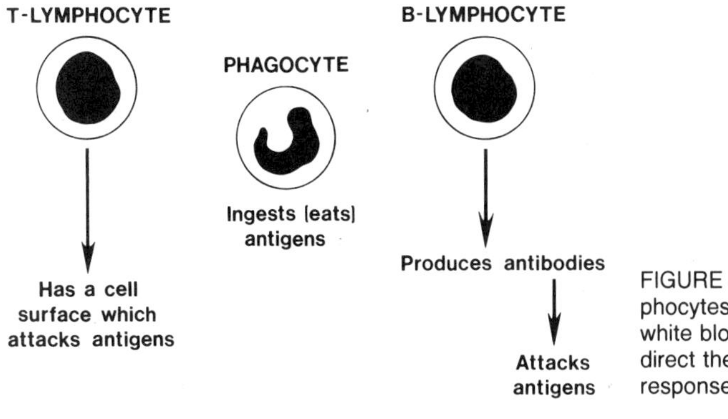

FIGURE 8-7 Lymphocytes: Specialized white blood cells that direct the immune response.

get it and some of us don't. In the last analysis, it is probably true that each cancer in our bodies is the result of a special set of circumstances including:

1. *the genetically determined uniqueness of the individual*
2. *the viruses we may harbor in our bodies*
3. *the unique pattern of our exposure to relevant environmental factors, including chemicals, cigarette smoking, and short-wave radiation.*

The underlying mechanisms that transform normal cells into cancer cells are essentially unknown. We cannot pinpoint, at the present time, the exact way in which the control mechanisms of normal cells are set aside by cancer cells. Nevertheless, approaches to the understanding of cancer through the cell cycle, DNA and protein synthesis, cell differentiation, and the immune response have been valuable. They have led to the development of cancer treatments based on the nature of the disease as it occurs in the various tissues of our bodies.

Types of Cancer

Any cell that is capable of dividing can become a cancer cell. Tumors, the products of unrestrained cell growth, can be benign or malignant. If they are benign, they remain localized and have a limited growth; if malignant, they do not remain localized but eventually metastasize, and their growth is typically unrestrained. Biopsies of tissues in which tumors have grown are processed and studied to determine whether the tumor is benign or malignant. Perhaps some of us can recall the waves of relief that flowed over us when a suspected growth was reported as benign.

Malignant tumors are derived from different types of tissues (see Table 8-1). About 85 percent of all tumors form in epithelial tissues, a type of tissue that is characterized by active mitosis (cell division). These tissues form covering and lining structures, and they also form glands in the body. Cancers in epithelial cells are called carcinomas, whereas connective tissue cancers (2 percent of cancers) are referred to as sarcomas. Leukemias are cancers that involve white blood cells; they constitute about 7 percent of all cancers in women. Pathologists have formulated detailed descriptions of tumors, based on refined description of their origin and structure; however, if we keep in mind the basic types of tumors, as outlined in Table

TABLE 8-1 Names and Tissue Origins of Some Human Cancers

Name	*Originating Tissue*
Squamous cell carcinoma	Skin
Melanoma	Skin (melanocytes)
Neuroblastoma	Nerve cells
Adenocarcinoma	Glands
Leukemia	White blood cells
Sarcoma	
Fibrosarcoma	Connective tissue
Myosarcoma	Muscle
Osteogenic sarcoma	Bone

8-1, we can have some idea about where tumors are and what kinds of tissues form them.

Cancer in History

Perhaps we think that cancer study and research, like the development of contraceptive technology, is a preoccupation of modern man. However, an examination of medical records from the pre-Christian era shows us that cancers, under different names, were already described in Egypt as early as 1500 B.C. Hippocrates (500 B.C.) gave the name "cancer" to these diseases. The various human tumors (from Latin *tumor* 'swelling') were classified by Galen before A.D. 200. Table 8-2 summarizes the historical development of the cancer concept and shows that the understanding of cancer has been linked to the understanding of structure and function in the normal cell. Both cancers and cells have been studied by scientists and phy-

TABLE 8-2 The Cancer Concept in Biology and Medicine

Year	*Event*
1500 B.C.	Cancer described in Ebers papyrus, Egypt
500 B.C.	"Cancer" named by Hippocrates
A.D. 164	Tumors classified by Galen
1775	Environmental cancers found in England
1790s	Cancer described as a tissue
1838	The Cell Theory: All living things come from cells
1869–Present	Cancer and the development of modern genetics
1910	Tumor cells transplanted
1910–1918	Cancers produced by 1. short-wave irradiation 2. viruses (?) 3. chemicals

sicians for a long time. In developed countries, at the present time, cancer has received more attention because more people live to die of it. Deaths from microbial plagues, diseases, wars, and famines are not common in affluent countries. On the other hand, living in relative affluence may predispose us to develop certain cancers, such as breast cancer.

CANCERS IN OUR BODIES

Breast Cancer

The archetypal and prototypical cancer for women is breast cancer. It is an ever-present threat. To a woman, breast cancer may mean the loss of a structure associated with her femininity, and to her and her family it may mean the disruption of family life, and death. There are few of us who have not had a personal encounter with this cancer through a mother, sister, relative, or friend. It comes as no surprise that it is the most common female tumor, accounting for 25 percent of newly diagnosed malignancies in women. Unlike most cancers, breast cancer can occur early in life; it is of major concern to the active woman between 30 and 70 years of age. [*Rubin, 1978*]

What are the chances of any one of us developing breast cancer, and why does it develop in some of us and not in others? The first part of the question can be answered from studies of the epidemiology of breast cancer. Epidemiologists chart the occurrence of breast cancer, and they can tell us about women who are most likely to have this disease. Women who have had breast cancer seem to have the characteristics listed below. If we ourselves have any of these characteristics, then the chance that we will develop breast cancer becomes greater:

1. *age over 35, especially over 50*
2. *family history in which a mother or sister had breast cancer*
3. *no children or children born after age 30*
4. *previous breast cancer in the other breast*
5. *chronic psychological stress.*

Other factors that may characterize the breast cancer patient include: high intake of dietary fat, living in the Western hemisphere in a cold climate, belonging to an upper socioeconomic group, and being white. Whether or not we have breast-fed our babies appears to have no relationship to breast cancer.

Obesity is also a predisposing condition for breast cancer. Greater body mass has been associated with earlier menarche, later menopause, and a longer reproductive life containing a larger number of menstrual cycles (Wallace et al., 1978). Prolonged exposure to estrogen seems to be implicated in these latter cases. It may be that some genetically transmitted endocrine factor directs irregular metabolism of steroid (particularly estrogen) hormones, and eventually causes breast cancer; however; at this time the etiology (cause-and-effect relationships) of breast cancer is not known.

Our breasts (Fig. 8-8) are made up chiefly of mammary glands, which are epithelial tissues, the kind in which 85 percent of tumors develop. Ligaments support the mammary glands (breast tissue), which are situated on top of the pectoral muscles. Fatty tissue on top of the mammary glands defines the size and shape of our breasts. Near the breast the axillary lymph nodes drain intercellular fluid from the tissues and return it to the general circulation. Breast cancers can involve only a small part of the breast, or all of it, and may invade parts of the surrounding tissue as well. If the cancer spreads to the lymph nodes, it can and does metastasize. Obviously, it is in our best interests to find the malignant tumor before it gets that far. How do we do it?

DETECTION OF BREAST CANCER

1. Breast Self-Examination (BSE). Roughly 90 percent of breast tumors are found by us (with possible help from husbands and lovers). A systematic and thorough breast examination, like a Pap smear, can literally save our lives. Figure 8-9 reproduces part of a pamphlet distributed by the American Cancer Society. We are advised to examine our breasts monthly, about a week after our menstrual period if we are still menstruating. If we are in the menopausal age group,

FIGURE 8-8 Anatomy of the normal breast and location of lymph nodes.

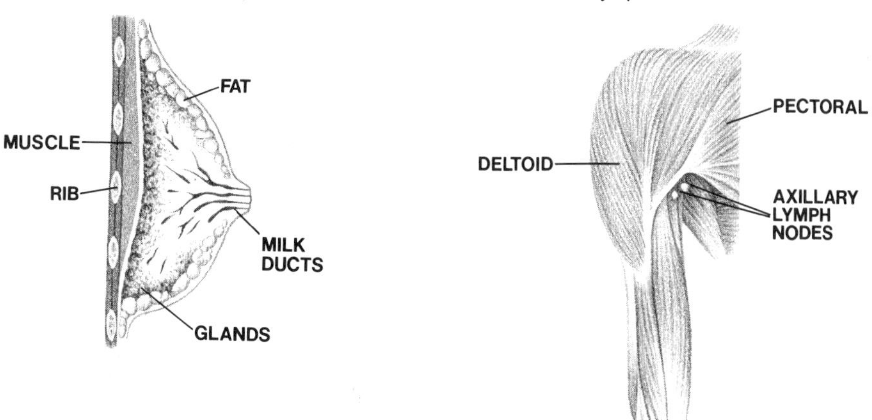

In the shower:

Examine your breasts during bath or shower; hands glide easier over wet skin. Fingers flat, move gently over every part of each breast. Use right hand to examine left breast, left hand for right breast. Check for any lump, hard knot or thickening.

Before a mirror:

Inspect your breasts with arms at your sides. Next, raise your arms high overhead. Look for any changes in contour of each breast, a swelling, dimpling of skin or changes in the nipple.

Then, rest palms on hips and press down firmly to flex your chest muscles. Left and right breast will not exactly match—few women's breasts do.

Regular inspection shows what is normal for you and will give you confidence in your examination.

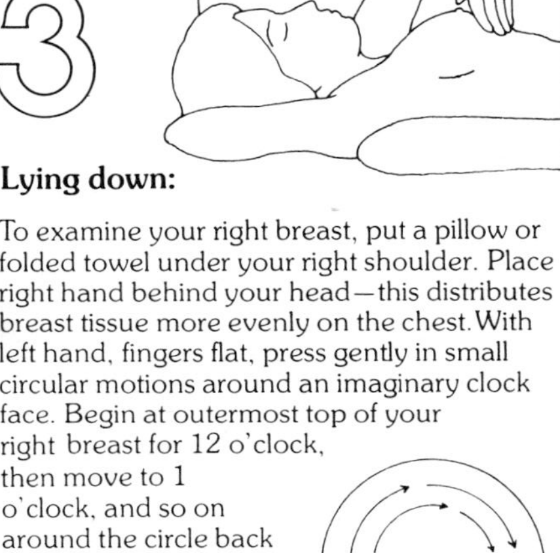

Lying down:

To examine your right breast, put a pillow or folded towel under your right shoulder. Place right hand behind your head—this distributes breast tissue more evenly on the chest. With left hand, fingers flat, press gently in small circular motions around an imaginary clock face. Begin at outermost top of your right breast for 12 o'clock, then move to 1 o'clock, and so on around the circle back to 12. A ridge of firm tissue in the lower curve of each breast is normal. Then move in an inch, toward the nipple, keep circling to examine *every part of your breast*, including nipple. This requires at least three more circles. Now slowly repeat procedure on your left breast with a pillow under your left shoulder and left hand behind head. Notice how your breast structure feels.

Finally, squeeze the nipple of each breast gently between thumb and index finger. Any discharge, clear or bloody, should be reported to your doctor immediately.

FIGURE 8-9 Breast self-examination. From "How to Examine Your Breasts" (New York: American Cancer Society, Inc., 1975), pp. 2–3. Courtesy of the American Cancer Society.

or have had a hysterectomy, we should make ourselves a consistent monthly time schedule.

It takes more than reading a pamphlet to be able to perform a good BSE. If possible, we should get some assistance, and familiarize ourselves thoroughly with the contours of our *normal* breasts. It is only then that we can recognize an abnormality if we feel it.

Something that feels abnormal to us may not be a cancerous condition. In women under 30, most lumps turn out to be benign nodules which may measure 1 or 2 inches in diameter. These should be watched carefully and examined by a physician, but they do not usually become malignant. Women over 30 commonly find cysts (fluid-filled sacs) in their breasts. These may be drained to remove the fluid. Most cysts disappear, particularly after the menopause, but women with cysts in their breasts should have regular examinations to be certain that no malignancies occur. It should be comforting to us to know that 90 percent of all suspicious lumps prove harmless.

2. Mammography. This type of breast examination uses x-rays to detect breast cancers before they are large enough to be felt during a BSE. Figure 8-10 shows the machine that delivers x-rays during a mammogram. The machine has been adapted during the last ten years to deliver lower dosages and thus to minimize the radiation risk to our other organs. (Note the lead cover on the patient's lap.) However, most physicians will not use mammography to examine all their patients for breast cancer because of the small, but finite, risk of radiation exposure. Guidelines suggested by the American Cancer Society are generally adhered to:

1. *All women should have breast x-rays once between age 35 and 40 to establish a reference, then at the advice of a physician until 50.*
2. *All women over 50 should have mammograms annually.*
3. *Women who have a family history (mother or sister with breast cancer) or a personal history of breast cancer should have mammography following the advice of their physician.*

3. Thermography. Thermography is a method of breast cancer detection that uses infrared rays rather than x-rays. Infrared rays are longer forms of electromagnetic radiation, and therefore have less energy which can disrupt biological systems. While it seems to be an attractive alternative to mammography, thermography is not reliable enough, at present, to be a diagnostic aid for breast cancer. One study has indicated that a combined use of clinical examination and thermography would have missed 14 percent of the cancers on initial examination (Strax, 1978). New adaptations of this technique may be more successful.

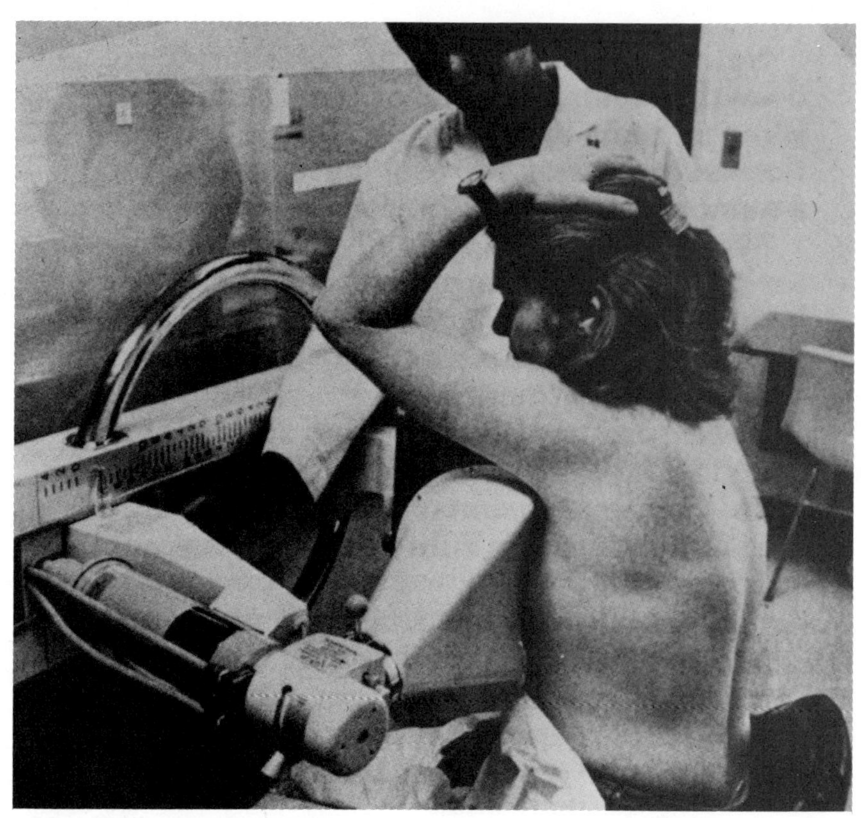

FIGURE 8-10 Having a mammogram taken. From "Facts on Breast Cancer" (New York: American Cancer Society, 1979), p. 2. Courtesy of the American Cancer Society.

4. Breast Cancer Screening Centers. Breast Cancer Screening Centers were established at 27 places in our country in order to measure the value of patient history, physical examination, mammography, and thermography—singly and in combination—in detecting breast cancer in women with no obvious symptoms. Each of these facilities, located as shown in the map in Figure 8-11, was funded jointly by the National Cancer Society and the American Cancer Society to provide examination for 10,000 asymptomatic women and to rescreen them annually for five years. These facilities have completed their screening tasks on schedule and the data are being assessed (Avery, 1980). We should be alert for media reports of the results.

5. Biopsy. If a BSE, mammogram, or thermogram indicates a possible tumor, a piece of suspected breast tissue will be removed surgically, then examined by a pathologist. This is the only way the physician can know for certain whether we have cancer cells growing in our breasts. We should remember that 90 percent of suspected breast tumors are benign, not malignant.

FIGURE 8-11 Breast cancer screening centers are located in each of the 27 cities shown on the map.

If malignant, then what? If the report comes to us, as it did to Marvella Bayh, that the tissue is malignant, we should be aware at this point that we have some choices. We can get opinions from another physician, preferably an oncologist (cancer specialist) who has had extensive experience with breast cancer. If any friends or acquaintances have had similar experiences, we should ask them for help in locating a well-qualified surgeon. "Dealing with a surgeon in whose judgment you have great confidence, is an indispensable first step in breast cancer treatment" (Rapaport, 1978).

TREATMENT Surgery may be recommended if the cancer is localized in the breast area. Again, early diagnosis can change the extent of the required surgery. For example, it may be necessary to remove only a small localized cancerous mass with some adjacent tissue (lumpectomy, tylectomy). On the other hand, removal of all the mammary gland (total mastectomy) may be required to be certain that all malignant tissue is taken from the body. A modified radical mastectomy will take the mammary glands with adjacent lymph nodes, whereas a radical mastectomy will involve the underlying pectoral muscles as well. Some controversy among physicians concerns the extent of surgery compatible with life and health in the cancer patient. Sometimes the cancer has spread beyond the point where surgery can help. Chemotherapy and radiotherapy, or various combinations may be suggested remedies. Their mode of action and side effects will be indicated later.

If we do, indeed, have to undergo a mastectomy, we may find that losing a part of our bodies that we have long associated with our femininity can be a devastating experience. We need all the help we can get at this time. Usually the person who can help us the most is another woman who has been through the same operation. She can assist us with adjustments in our self-image, in our concepts of our sexuality and femininity and in our family's acceptance of what has happened to us, because she has faced the same adjustments herself.

Marvella Bayh described her postoperative contact with a representative of the Reach to Recovery program for breast cancer patients:

As I was feeling down, stringy-haired and unattractive after a week in the hospital, into my room walked a woman looking like a million dollars, in a jersey blouse, slit midi-skirt and stunning hair-do. I was tempted to pull the covers over my head. "I'm Virginia Newman, from Reach to Recovery," she said. "And I want to talk to you." As a volunteer from the American Cancer Society, her job was to show me how to wear a prosthesis (artificial breast), to exercise my arm with a ball and rope, which she brought along in her kit, and, most important, to make me feel like a whole woman again.

Virginia Newman, with her sleek figure, looked me in the eye and said, "A few years ago, I had a breast removed, just like you. You must remember that you are the same person today that you were before surgery—in every way." She talked with me for more than an hour, while I looked her up and down, deciding that I'd better get with it and look as smart as this woman did. She gave me advice that meant more from her than it would have from any doctor, unless the doctor were a woman who had also had a mastectomy. [Bayh and Kotz, 1979]

The Reach to Recovery program, in its entirety, is described in a publication of the American Cancer Society written by Terese Lasser (1974). The booklet describes this program, a free service of the American Cancer Society. It is available, with her physician's approval, to any woman who has had breast cancer surgery.

Treatment for breast cancer may include radiation therapy and/or chemotherapy. These treatments are also prescribed for other cancers in our bodies, and they will be described in the section on dealing with cancer.

The importance of early detection of breast cancer is shown by statistics on five-year survival rates. Of women whose breast cancer is located before it has spread to the axillary lymph nodes, over 80 percent are alive and well five years after surgery. On the other hand, if the cancer has extended beyond the breast, only about 50 percent remain alive and well at the five-year point. Fear of the unknown may prevent our seeking help when we find that we have some sort of abnormality in our breasts, but the longer we wait to have something done about it, the shorter our survival period, if, indeed, we do have cancer.

We can take charge of our own health and well-being insofar as breast cancer is concerned by following these "Ten Precepts for Protection":

1. *Remember that breast cancer is a curable disease—if caught in time.*
2. *Note that we have the methods and equipment to detect breast cancer in its early stages when it is still confined to the breast and is therefore curable.*
3. *Be aware of the enemy and realize that we have powerful weapons on our side.*
4. *Our most valuable weapon is our own hands. We should use them properly and regularly once a month in doing breast self-examination.*
5. *Visit our physician once a year—or more frequently if he or she finds it advisable—for a complete examination. Such an*

examination may require more equipment and expertise than our physician has. If so, ask for special studies to be arranged.

6. *Remember that most breast conditions are harmless. At the same time, remember that we owe it to ourselves to inform our physician of any new findings in our breasts, for only he or she can make the proper differential diagnosis.*
7. *Alert all women close to us to the above advice, and urge them to follow through.*
8. *Learn as much as we can about our breasts and their disorders. Ask questions of our physician if we do not understand.* The more knowledge we have, the better we can protect ourselves. *Keep informed about new discoveries—we never know when they may be useful.*
9. *Never take lightly a change in our breasts. The reassurance of a negative examination by our physician is worth more than the effort involved in getting it.*
10. *Above all, remember that the odds are strongly against our ever developing breast cancer—but that we must be on our guard at all times.*[2]

Gynecological Cancers

Cancers (of all kinds) are the leading cause of death among women who are 30 to 54 years old. Breast cancer, as we have seen, is the most common cancer in women, followed in frequency by gastrointestinal cancer. Gynecological cancers, which include cancers of the uterus, ovaries, and vagina, among others, are now third among our cancer enemies.

During the last 50 years the death rate from uterine cancer has decreased steadily, while that from ovarian cancer has increased slightly. Death rates from these and other types of cancer are depicted in Figure 8-12. Early diagnosis of cancer can help us avoid becoming one of these statistics.

CERVICAL CANCER The decrease in uterine cancer as a cause of death among women during the past 50 years has been directly related to the diagnostic procedure for cervical cancer developed by Dr. George N. Papanicolaou. Papanicolaou studied vaginal cytology and developed techniques that showed cyclical changes in the vaginal cells as well as cancer-caused abnormalities in these same cells. His

[2]"What Else Can You Do: Ten Precepts for Protection" (pp. 85–86) in *Early Detection* by Philip Strax, M.D. Copyright © 1974 by Philip Strax. Reprinted by permission of Harper & Row, Publishers, Inc.

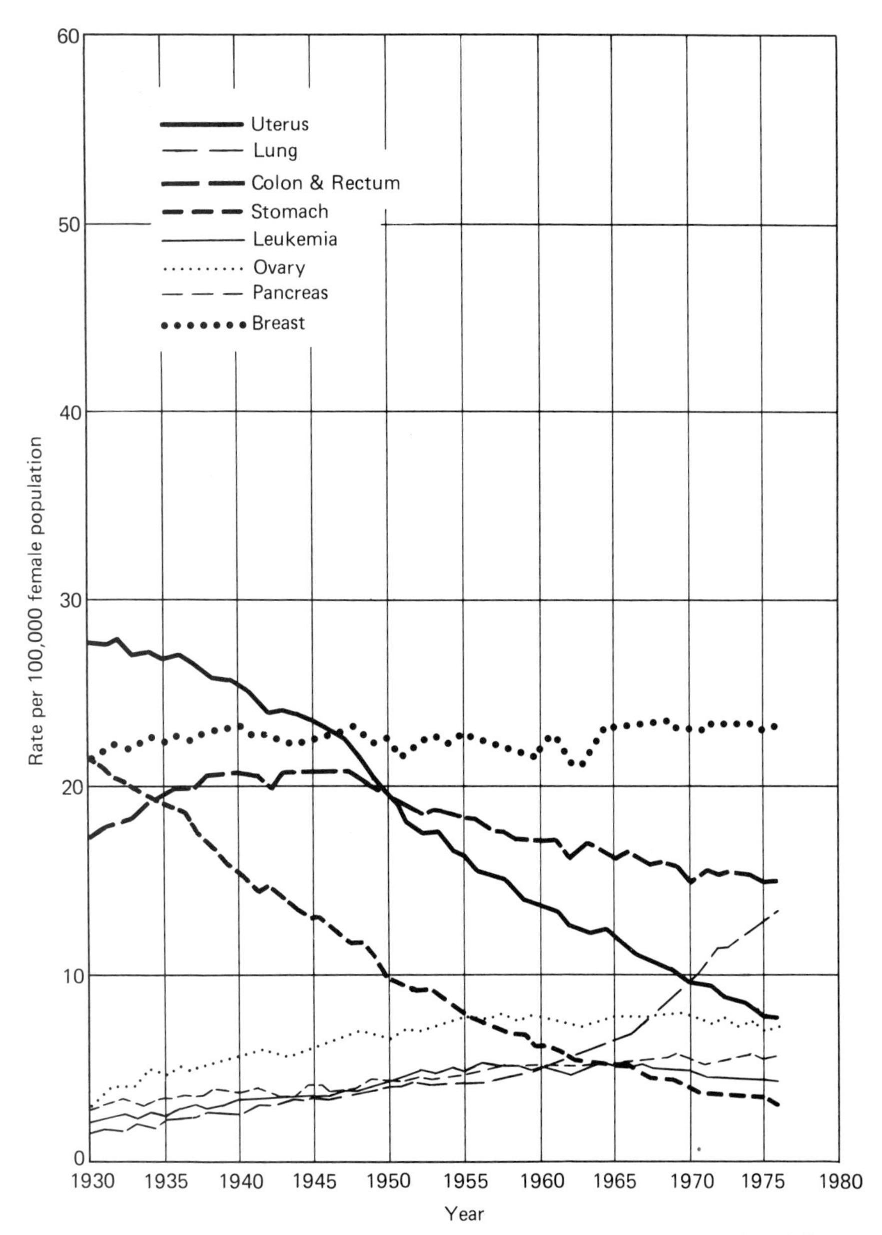

FIGURE 8-12 Age adjusted cancer death rates for selected sites, females, United States, 1930–1976. Death rates are standardized on the age distribution of the 1940 U.S. Census Population; sources of data are U.S. National Center for Health Statistics and U.S. Bureau of the Census. From Lawrence Garfinkel and Edwin Silverberry, "Cancer Statistics, 1979." New York: American Cancer Society, 1979. Courtesy of the American Cancer Society.

pioneering cytological studies have saved the lives of thousands of women.

We may recall our doctor taking a "Pap smear" during our annual checkup. Although the prone, feet-in-stirrups lithotomy position, illustrated in Figure 8-17(D), is not a very good one from which to make accurate observations, we may have been aware that the doctor took some cells from the cervix or neck of the uterus and vagina (Fig. 8-13, top) and placed these cells on a slide. The slides were processed and "read" by a cytotechnologist. The cells on the slide originated in the cervix and vagina and were exfoliated (shed) as they were replaced by cells growing in deeper layers of the cervical and vaginal epithelium. Abnormal cancerous or precancerous cells in the preparation have characteristics that can be detected by the trained observer.

In most cases, the report we received from the doctor's office or laboratory was negative. Those of us who had reports that our Pap smear was suspicious or positive were asked to have a repeat examination. During this examination, the physician took some tissue from various parts of the cervix itself for a biopsy (Fig. 8-13, bottom) and examined the cervix with a special microscope called a colposcope. The cervical epithelium observed with either system was either atypical (different from other uterine tissue but not malignant) or malignant. Untreated cervical cancer can metastasize to the lymph nodes, vagina, and endometrium, as well as to other parts of the uterus. Rarely, it can metastasize to different sites in the body by way of the circulation.

Other than the abnormalities in the cells of the cervix and vagina, detectable by the Pap smear, symptoms of cervical cancer include: unusual vaginal bleeding, particularly after intercourse, an unusual vaginal discharge, and an abnormally heavy menstrual flow or spotting between periods. Cells in either the upper or lower layer of our cervical epithelium can become cancer cells, acquire their own blood supply, and disrupt the normal protective and secretory function of the cervix.

Cervical cancer is more common among women who have had frequent sexual experience (1) in their early teens, and (2) with many different partners. Is cancer of the cervix then, a sexually transmitted disease? Not in the usual sense. A study of Taiwanese prostitutes (Sebastian et al., 1978) who did not begin their sexual activity until after age 18 showed that coitus per se was not the causative factor in the disease, but that the timing of the activity was important. These women, tested for cervical cancer during their legally prescribed venereal disease examination, showed only a 1 percent incidence of cervical cancer. The two variables, time and multiple partners, were separated in the Taiwanese women. Apparently the sexual activity that might result in cervical cancer is confined to the period of early adolescence or to the first pregnancy when the

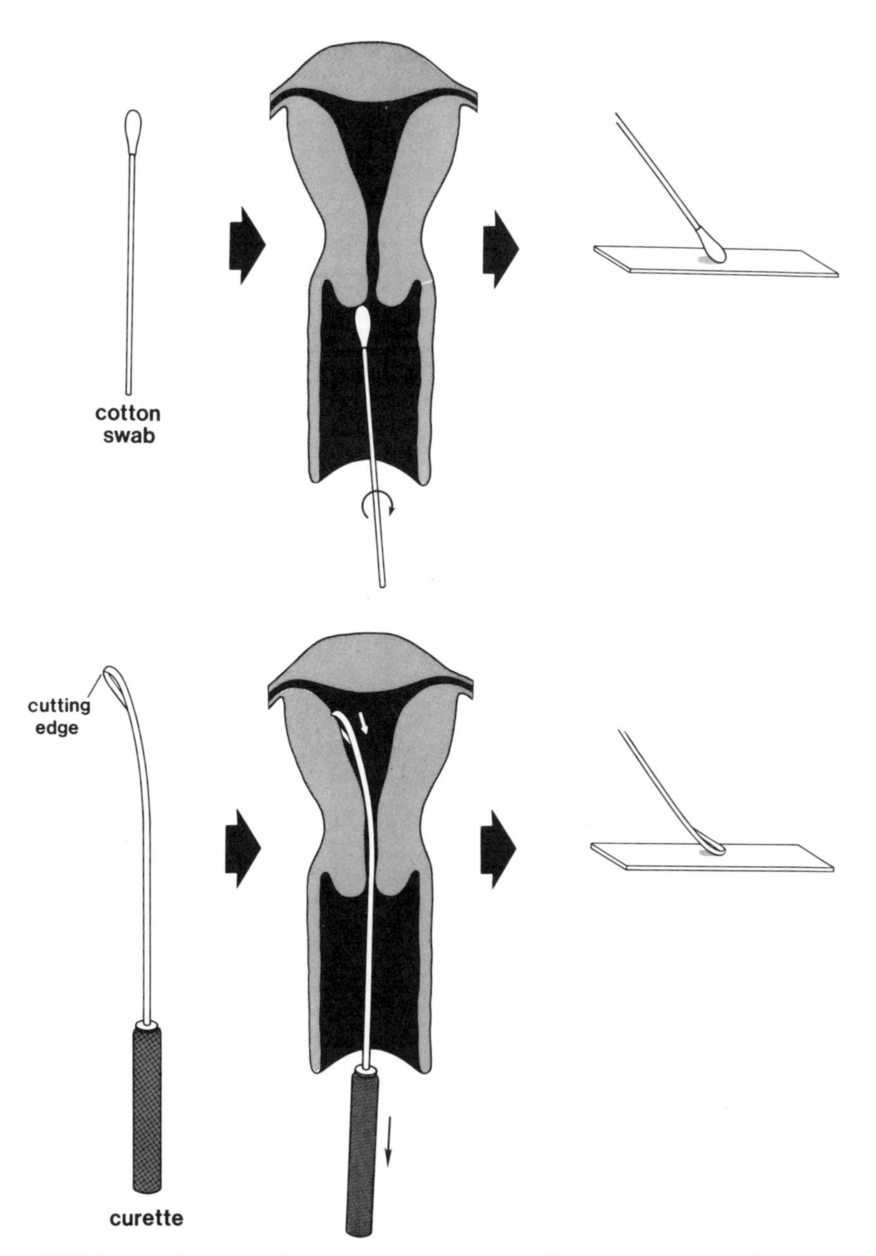

FIGURE 8-13 Endometrial biopsy and cervical smear. The material obtained during the endometrial biopsy is processed before it is put on the slide.

epithelium of the cervix is undergoing its most rapid growth and proliferation. At this time the cervix is possibly more vulnerable to external influences that may program genes in some cervical cells to develop into the malignant state.

Other epidemiological factors connected with cervical cancer include low socioeconomic status and poor care during or following a pregnancy. The incidence of cervical cancer is highest in women 40 to 49 years of age. The long latent period between exposure to the disease and occurrence of symptoms could indicate that a virus, or some other coitally transmittable antigen, is implicated in cervical cancer. Genital herpes virus has been suggested as a causative agent for cervical cancer. If this is the case, it should be possible to isolate the virus from malignant tissue and/or to demonstrate antibodies to the virus in the serum or plasma of women who have or who have had cervical cancer.

While research into the cause of cervical cancer continues, we should keep in mind the weapons for defense that we presently have. Unusual vaginal bleeding is a warning sign that we should heed. Currently, the American Cancer Society recommends that women aged 20 to 65 and sexually active women under age 20 need Pap smears only every three years, *if* two tests done a year apart are negative. We should follow the advice of our own physician about the frequency of our Pap smears. If we are over 40, we need to exercise increased vigilance, because we are coming to an age when endometrial cancer could also be a part of our lives.

ENDOMETRIAL CANCER The endometrium, the lining of the main part (body) of the uterus, is the site for implantation of the embryo during normal pregnancy. It is also a possible site for the growth of cancer cells. Active growth and proliferation characterize the endometrium, particularly during our reproductive lives when it is shed cyclically during the nonpregnant ovulatory cycle.

Unfortunately, the Pap smear is of limited value in the diagnosis of endometrial cancer, since few of the endometrial cells are normally shed into the cervix and vagina. A specimen of the endometrium must be obtained by curettage (scraping), illustrated in Figure 8-13(B), or by aspiration (removal under pressure). The specimen requires a histological investigation, which includes preparation and microscopic study of the uterine tissue.

External indications of endometrial cancer include vaginal bleeding in postmenopausal women, or abnormal menstrual flow and spotting between periods in premenopausal women. Both groups may experience vaginal bleeding after intercourse; these symptoms of endometrial cancer are similar to those of cervical cancer. The diagnosis and treatment patterns are similar.

Cancer of the endometrium is a disease of mature women. Fourteen percent of us who are over 35 will develop it eventually, most frequently when we are in the 50 to 64 age bracket. We are considered to be at high risk for endometrial cancer if we have personal medical histories that include diabetes, high blood pressure, or obesity. Reproductive disorders that predispose us toward developing endometrial cancer include failure to ovulate and subsequent infertility, late menopause (after age 55), and irregular menstrual cycles. Abnormal uterine bleeding that began during puberty may be an indication that we are more likely to be in the endometrial cancer group. Those of us who have never had children or who have had our children later in life (after age 30) are considered to be at higher than average risk for developing endometrial cancer. Menopausal women, and those who are close to that stage in life, are more likely than others to develop cancer of the endometrium. If we take supplementary estrogens for the relief of our menopausal symptoms, we increase the likelihood of endometrial cancer in our bodies.

Women who spend most of their lives under the influence of estrogen (due to continuous menstrual cycles, uninterrupted by pregnancy, or to anovulatory menstrual cycles) seem to be more prone to develop endometrial cancer. Thus far, the reason for this is unclear. It may be related to the general way that steroid hormones affect the chromosomal-genetic apparatus of cells (Fig. 1-7). In this regard, it is of some interest to note that progesterone, the other hormone secreted by the ovary, has been used to treat uterine cancer. Progesterone-like substances (progestogens) seem to act on responsive endometrial cells in much the same way as progesterone acts on the endometrium during the ovulatory cycle, when it transforms an endometrium that is growing without differentiating into one that is nondividing, but specialized to form glands and to secrete their products (Reel, 1976).

The knowledge on our part that we belong to a group of women who are more likely than most to develop endometrial cancer should lead to increased watchfulness for external symptoms. We should also request an endometrial biopsy (histological examination) during our annual or semiannual gynecological examination, particularly if we are taking estrogen supplements.

Ovarian Cancer Ovarian cancer, the fifth leading cause of cancer deaths in women, is lethal because it is almost never found until it is too late to do anything about it. Ovarian tumors can grow to be very large; an extreme example was a 200 pound ovarian tumor recently removed from a 24-year-old woman in San Francisco. Although ovarian tumors weighing more than 20 pounds are rare, they do not have to be that large to be dangerous to our health. The 30 to 35 percent survival rate for ovarian cancer has remained essentially unchanged for 30 years. Survival time (after surgery for ovarian cancer) varies with age from 5.4 years in women under 45 to less than a year in women over 65. About 200,000 women per year die of ovarian cancer.

Early detection of ovarian cancer, with all its life-prolonging ramifications, is largely a matter of chance. During a pelvic examination the physician palpates (feels) our ovaries. Since the size of our ovaries varies with our reproductive state (they become smaller with age and the menopause), the presence of a 25-year-old ovary in a 60-year-old body may indicate the presence of an ovarian tumor. Painless and rapid increase in body size particularly around the waist and abdomen may be a late symptom of ovarian cancer.

Most ovarian cancer cells grow in the outer covering of the ovary rather than in the inner part. They can, therefore, break away from the ovary and spread to other organs in the pelvic cavity. Unfortunately, by the time ovarian cancers are detected, many of them have reached this stage, and therefore surgery for ovarian cancer may involve the removal of other structures as well. Radiation therapy and/or chemotherapy may be adjunct treatments if cancer cells have metastasized to organs where surgery would be contraindicated.

The peak incidence of ovarian cancer occurs in women between the ages of 50 and 59. Women in this age group appear to be at added risk if they have experienced increased premenstrual tension, heavy menstrual periods with increased breast tenderness, a tendency toward spontaneous abortion, infertility, an early menopause, or have never borne children. Women with cancer of the breast have twice the expected risk of developing ovarian cancer, whereas women with cancer of the ovary have three to five times the expected probability of developing breast cancer. Early pregnancy apparently gives some protection against ovarian cancer as well as against endometrial cancer.

It would seem that breast, endometrial, and ovarian cancers can develop in women who may have a genetically directed variation in levels of hormones secreted by the hypothalamus, pituitary, and ovary respectively. These women may also have hereditary defects in the enzyme-directed pathways of hormone interconversion. Estrogen receptors, by their presence or absence, may be implicated in tumor development.

At the present time, because the early detection of ovarian cancer is not possible, and because the detection of ovarian tumors even in more advanced stages is largely a matter of chance, we can decrease our chances of having advanced ovarian cancer by heeding the general danger signs of all cancers:

C hange in bowel or bladder habits

A sore that does not heal

U nusual bleeding or vaginal discharge

T hickening or lump in breast or elsewhere

I ndigestion or difficulty in swallowing

O bvious change in a wart or mole

N agging cough or hoarseness.

Vaginal Cancer Vaginal cancer is rare in women under 60. Vaginal discharge, bleeding, and pain would be some indications for diagnosing this disease in postmenopausal women. However, a group of much younger women has been identified in whom adenocarcinoma of the vagina was totally unexpected. A careful, case controlled, retrospective study (Herbst et al., 1971, 1975) of these young women, ages 15 to 22, indicated that their symptoms were associated with medication taken by their mothers to support a high-risk pregnancy. Because the mothers of these girls had difficulties carrying pregnancies to term, their physicians had prescribed hormonal assistance in the form of diethylstilbestrol (DES). Symptoms of vaginal abnormalities including adenosis (the presence of endometrial-like epithelium in the vagina) or adenocarcinoma (malignant tumors of the vaginal epithelium) were found in girls whose mothers had taken DES before the eighteenth week of their pregnancies. The ingestion of DES by the mothers has also been linked to reproductive difficulties, including a high fetal death rate, in these affected daughters. The reproductive difficulties appear to be more numerous than the cancer complications (Barnes et al., 1980).

The apparent relationship between maternal ingestion of DES and cancer and reproductive difficulties in female offspring separates a whole population of women from the general population. In a sense, this cause-and-possible-effect relationship is a model that might be applied some day to the detection of gynecological and nongynecological cancers in our bodies. One factor with a clear-cut etiology (cause-and-effect relationship) indicates that some of us should be watched more carefully than others for the early appearance of a specific form of cancer. Our physicians will examine us more thoroughly and frequently if they know that we are in this group of women at high risk for vaginal cancer. Such surveillance, economically unfeasible for all women, becomes practical and even necessary when it can be focused on a smaller, well-defined population with a known exposure to a causative agent (Ulfelder, 1976). If some one, simple etiological factor could also be identified for breast, uterine, and ovarian cancer, the mortality rate for these diseases might be decreased by the early identification of high-risk subjects.

Cancers Related to Pregnancy Cancers of all types have been reported during pregnancy or in the immediate postpartum period. No matter what the type or extent of the cancer, pregnancy itself introduces other complications to the diagnosis and treatment of cancer:

1. *Age. Younger women (except for the DES daughters above) are not considered to be at high risk for cancers, and are not usually examined for them.*

2. *The superimposed pregnancy makes diagnosis more difficult.*
3. *Adequate treatment for cancer may terminate an existing pregnancy.*

Some women may also develop trophoblastic tumors as a result of the pregnancy. These tumors, which may become malignant, originate from the part of the embryo that forms the placenta. Cells of the trophoblast may grow and multiply, then metastasize in the typical manner of cancer cells. As they increase in size and number, they produce a hormone that is characteristic of the pregnant state, human chorionic gonadotropin (HCG). As shown in Figure 8-14, the growth of the tumor is directly proportional to the amount of HCG it produces.

Although these gestational trophoblastic tumors are rare, we can look at them as a potential model for detecting cancer and for following its response to treatment. If each type of cancer cell also secreted a distinctive product, the product could be measured as an indication of the extent of tumor growth. Regression of the cancer following radiotherapy or chemotherapy could also be monitored externally. This same kind of relationship between the cancer and its product might also be demonstrated one day for cancers that occur in other parts of our bodies.

Cancers in Other Organs

We know that cancers can and do develop in other, nonreproductive parts of our bodies. Three of these cancers—lung cancer, colorectal cancers, and skin cancers—merit our attention. By taking relatively simple precautions we can protect ourselves from these diseases. We can also understand some of the risk-benefit relationships in-

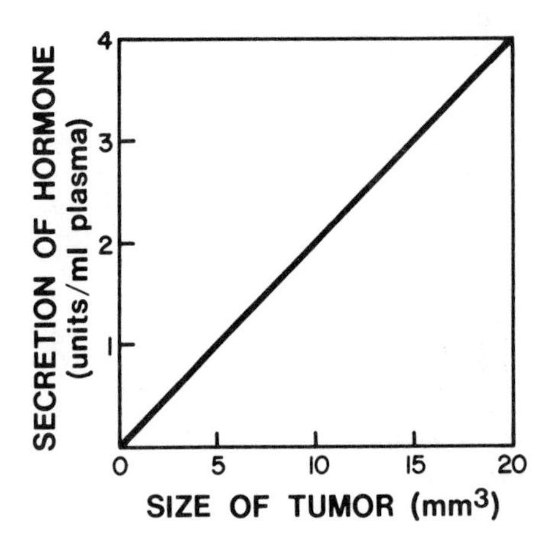

FIGURE 8-14 Following tumor growth by hormone measurement.

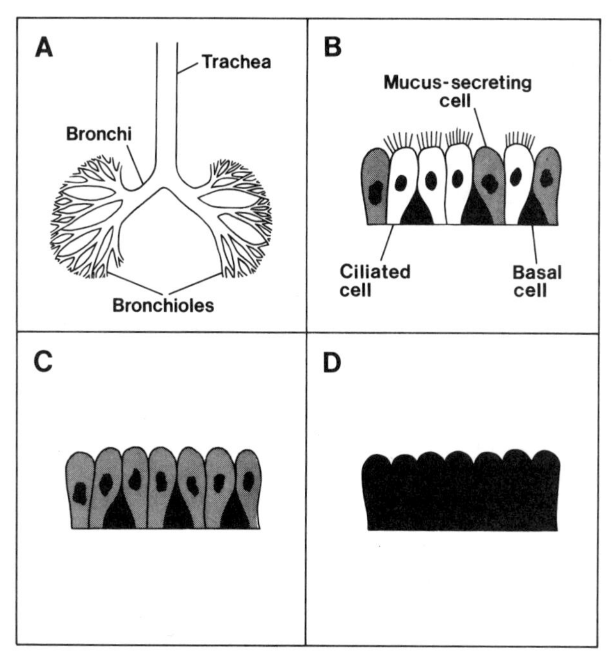

FIGURE 8-15 The respiratory tract and the cells which line it: Effects of smoking. (A) Structure of the respiratory system. (B) Ciliated epithelium which lines the structures in (A). In healthy epithelium, ciliated cells operate to move pollutants from the lungs. (C) Increase in mucus cells and loss of ciliated cells which may occur after smoking cigarettes for a period of time. (D) Induction, by carcinogens in cigarette smoke, of growth in basal cells. These cells become malignant. They replace the specialized cells of the epithelium and develop into tumors of the lung and/or other respiratory structures.

volved in such taken-for-granted activities such as cigarette smoking and suntanning.

LUNG CANCER Cancer cells develop most frequently in the epithelial cells which cover our bodies and line the inner organs of our digestive, urinary, respiratory, and reproductive systems. The epithelial lining of the respiratory tract, which includes the trachea, bronchi, and bronchioles [Fig. 8-15(A)], consists of several layers. Two kinds of cells, mucus-secreting cells and ciliated cells constitute the upper layer of the respiratory epithelium. Ciliated cells move polluting substances away from the lung alveoli, where the exchange of respiratory gases takes place. Lower layers of the epithelium consist of smaller cells, called basal cells, that divide rapidly and usually produce either ciliated cells or mucus cells. Figure 8-15 shows, diagrammatically, the mixed pattern of cells that line the normal, healthy respiratory system (B), and the degenerated cell patterns that result from exposure to cigarette smoke and other pollutants (C) and (D).

Both smoking and the constant exposure to environmental pollutants stress the cells of the respiratory epithelium in two ways. First, the ciliated cells have too much to do. They can no longer move pollutants up toward the throat and mouth because they are overwhelmed by the polluting substance. They die and are replaced by mucus-secreting cells. Pollutants and mucus have no way to get out of the respiratory tract; coughing is the smoker's way of reacting to the presence of pollutants in her lungs. The chemicals in cigarettes or in the environment may also turn on the latent cancer potential in the basal cells. These cells proliferate rapidly in the normal state. They usually form the other cells, but, if changed into cancer cells, they replace both the ciliated and mucus-producing cells. Their metastases are aided by the very rich blood supply to the lungs. The presence of large clumps of cancerous or precancerous cells obstructs the bronchi and bronchioles. A reflex coughing reaction to the obstruction is a warning signal of their presence.

As Figure 8-12 indicated, lung cancer in women is on a rapid upswing. Cancer authorities estimate that, by the middle of the 1980s, lung cancer will approach breast cancer as a leading cause of death among American women. This increase in the incidence of lung cancer in women has been linked to the movement of women into the higher, stress-related echelons of the job market. Cigarette smoking may be a mechanism used by women to deal with these stresses.

Whatever our reasons for smoking cigarettes, this practice has a deleterious effect on our own and our children's health and well-being, and it increases the chances that we will develop lung cancer. The most common early symptom of lung cancer is a cough. A *lingering* cough, combined with blood in the sputum, should send us to our physician without fail. Late in the disease, other symptoms such as chest pains, loss of weight, difficulty in swallowing, and the accumulation of fluid in the chest cavity will also convince us that something is wrong.

The detection of lung cancer is generally made by x-rays, but there are some lung tumors that x-rays will not isolate in early stages. Cytological examination of cells in the sputum may reveal the presence of cancer cells in the lung. At times, the physician may perform a biopsy of the lung tissue itself.

The prognosis for lung cancer is so bad that the best defense is not to get it at all. If the cancer is small enough to be removed by surgery, one-fourth of the lung cancer patients can profit by this treatment. However, only half of them will be alive five years after the operation. Chemotherapy can be used to prolong the life, briefly, of one type of lung cancer patient. Radiation therapy may alleviate some symptoms of the disease. Like ovarian cancer, lung cancer is lethal because, by the time it is detected, it has usually metastasized beyond control.

We don't know yet why some of us get ovarian cancer and

some of us don't, but we do know that circumstances within our control can minimize our risk of getting lung cancer. Some of us will get lung cancer and some of us won't, but most of us who won't don't smoke.

COLORECTAL CANCER

Educational programs have emphasized lung, breast and uterine cancer, placing less emphasis on colorectal cancer. Colorectal cancer, however, is the most common visceral cancer in humans and is exceeded in mortality only by lung cancer in men and breast cancer in women. The overall results of treatment have not improved over the past several decades because cancer of the large bowel is not being diagnosed in the early states. The large bowel is often investigated only if such symptoms as bleeding, change in bowel habits, abdominal pain, anemia, or loss of weight are present. Cancer, when found with such symptomatology, is usually advanced and the five-year survival rate is about 40 percent. On the other hand, when cancer is found on routine investigation in asymptomatic or mildly symptomatic patients, the five-year survival rate is increased to 88 percent. [*Abrahamson and Morton, 1978*]

Here is a cancer that can be easily detected, if only someone looks for it. The final segments of the large intestine, the sigmoid colon and the rectum (Fig. 8-16), are the parts of the intestinal tract from which water is reabsorbed and fecal material is excreted. Perhaps some of us think of this part of our body as "dirty" and feel an esthetic repugnance toward allowing someone to examine it. Tumors in this area can't be found, however, unless someone looks for them. Such tumors develop more frequently after we reach 40.

The epithelial lining of the large intestine is the site of active cell division; these areas are likely sites for tumors to begin developing. Most colorectal tumors are nonmalignant growths called polyps, which can be removed quite simply. Cancerous tumors usually appear close to the rectum. Because bleeding that could be associated with hemorrhoids is sometimes caused by cancer, any bleeding in this area should be reported promptly. Other signs of colorectal cancer include noticeable changes in bowel habits, diarrhea, constipation, and gas pains. In most cases, the cancer is well-developed before the symptoms appear.

The early stages of colorectal cancer, unlike most cancers, are relatively easy to detect. The physician can feel the presence of a tumor by a finger probe and see it by using a device such as a proctosigmoidoscope or other instrument for visualizing the colon. Of the four examination positions shown in Figure 8-17, (A), (B), or (C) may be used for rectal examinations. X-ray examination following a barium enema may also be used to diagnose the presence and extent of colorectal cancer. A chemical test of fecal material can be used to detect blood, and possible tumors of the large intestine.

Radiotherapy and chemotherapy are not effective treatments for cancer of the colon or rectum. Surgery is more effective (70 percent survival rate) when the cancer has not reached the lymph nodes. Extensive surgery may involve the reconstitution of the colon or the rerouting of fecal material through an artificially created opening (ostomy). Ostomy patients, who must adjust to new ways of carrying out some taken-for-granted body functions, have formed support groups similar to the Reach for Recovery programs for breast cancer patients. Survivors help survivors make life-style adjustments to the surgery that has probably saved their lives.

SKIN CANCER On the first warm days of spring, out come the bikinis, and we are off to the ritual pursuit of the summer tan. The exposure of our bodies to the sun's rays, with subsequent darkening of the skin, is somehow connected in our minds with the glow of health and all around physical fitness.

The warm rays of the sun, however, contain ultraviolet radiation, a form of energy which, as we have seen, can induce malignant changes in normal cells. One of the functions of our skin is to protect our bodies from ultraviolet radiation. This function is

FIGURE 8-16 Location of rectal cancer.

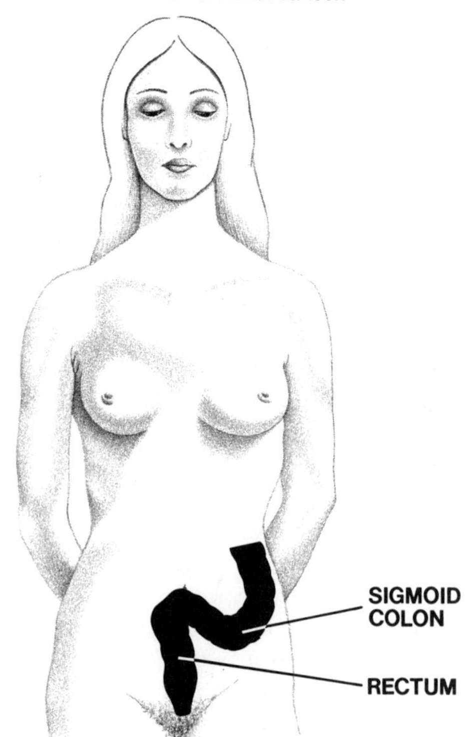

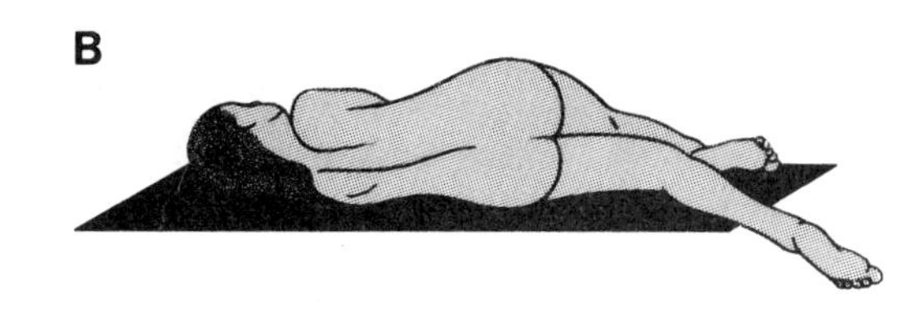

FIGURE 8-17 Positions for examinations. (A), (B), and (C) are positions commonly used for colorectal examination. The lithotomy position (D) is used for gynecological examination.

carried out by melanocytes, the pigment-forming cells in the lower part of the skin's epidermis.

This upper layer of our skin, the epidermis (Fig. 8-18) consists of several layers of cells which are derived from the lowest, basal layer, the site of active mitoses. The cells next to the basal layer derived from them are the squamous cells. Cells in the lower layers of the epidermis multiply and move up toward the surface. They form the outer, dead layer of skin. If we run our fingernails over our skin, the layer of dead cells comes off as a whitish mark.

Skin color depends upon, among other factors, the number of melanocytes present in the epidermis and their relative production of melanin pigment. This is a genetically determined characteristic; fair-skinned persons simply do not have as much melanin as darker, olive-skinned individuals.

Exposure to sunlight first darkens melanin pigment already present (Fig. 8-19). This condition is reversible. Increased exposure causes the melanocytes to produce more pigment in order to protect the skin. The fewer the melanocytes, the less protection against the sunburn-peel result of too much exposure with too few melanocytes.

Besides the painful consequences of a sunburn, we sun-worshippers can put ourselves into a high-risk group for skin cancer. This and lung cancer are the only cancers we've looked at so far in which membership in the high-risk group was a matter of choice.

Unprotected exposure to ultraviolet irradiation can induce the

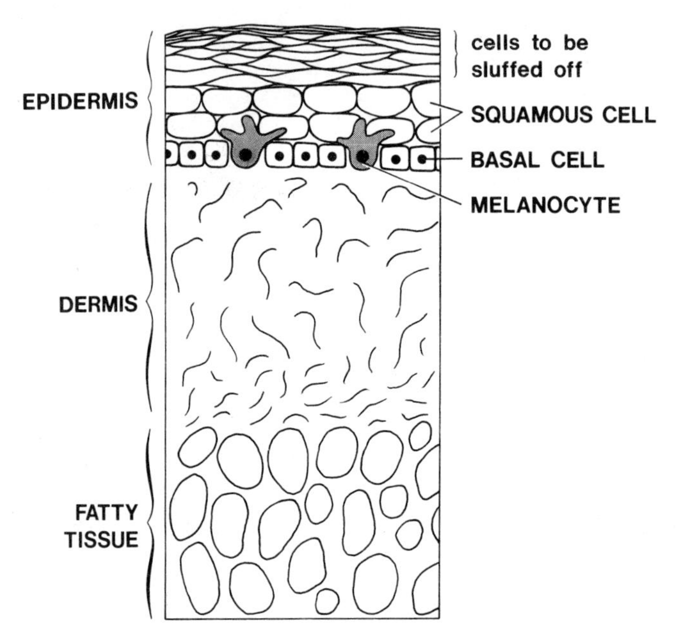

FIGURE 8-18 The layers of the skin.

basal and squamous cells to form nonmalignant tumors. When these tumors are less than one inch in diameter, their surgical removal results in a cure in nearly 100 percent of all patients. Their visibility facilitates their early detection and removal. If skin tumors are neglected, however, they may grow inward and interfere with underlying tissue and/or openings such as the eye, ear, or mouth. Squamous and basal cell carcinomas develop in fair-skinned persons whose exposure to the sun is cumulative, that is, who work outdoors in sunny climates.

Melanoma, cancer of the melanocytes, is a different disease. Almost all of us have freckles or moles, which represent excess concentrations of melanin or melanocytes respectively. What is the difference between melanoma and a mole? Here's an illustration:

Helen was feeling chipper and as a thirtieth birthday present to herself she had a medical checkup. In examining her upper back, the doctor

A MELANOCYTE WITH PIGMENT

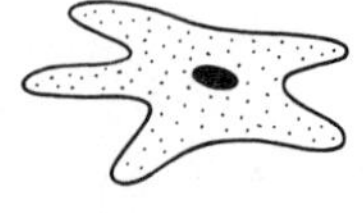

moderate exposure to sun

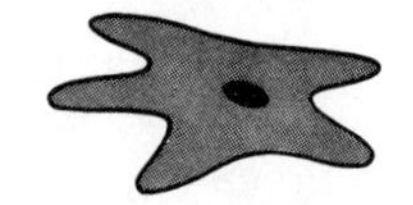

increased exposure to sun

FIGURE 8-19 Protection against ultraviolet radiation.

spotted a dark black mole with irregular borders that he thought ought to come off because it was "funny looking." Helen recalled that it had bled when her bra rubbed it and that it had grown "a little" recently.

The doctor suspected that she had a form of skin cancer called a melanoma, but the only way he could be sure was to remove it entirely and then have a pathologist examine it under a microscope.

But, by not preparing Helen's specimens correctly, the pathologist measured the extent of her cancer imprecisely. He had sliced the mole in such a way as to measure only the thinnest—and not the thickest—part. As a result Helen's doctor was misled into treating her less vigorously than should have been the case.

A few weeks later, Helen consulted a dermatologist. When he examined the specimens of the melanoma under the microscope, he recognized the other doctor's mistake. Helen received the appropriate therapy: more extensive surgery to remove a wider area around the original site of the mole and nearby lymph nodes as well as chemotherapy. Melanomas kill by spreading to and destroying the lungs, brain, and other vital organs. The aim was to thwart the cancer's growth. Despite the therapy Helen died at age 31.[3]

Melanomas are becoming more common in this country, particularly among younger people. They are seen most frequently in those of us whose exposure to the sun occurs in weekend or vacation bursts. Weekends or vacations, if they involve exposure to the sun, should include sunscreen lotions. As the distance from the equator decreases, our need for sunscreens increases if we want to protect our skin, and ourselves, from the painful, occasionally even fatal, results of indiscriminant sun exposure.

Cancer, Oral Contraceptive Steroids, and Estrogen Replacement Therapy

ORAL CONTRACEPTIVES Oral contraceptive steroids (OCS) have been in use by large numbers of women for less than 20 years. Since the development of cancer requires a long latent period in most cases (Hertz and Bailer, 1966) it is really too soon to say "Yes, OCS cause cancer," or "No, they don't."

As we have seen, the development of breast, uterine, and ovarian cancer usually occurs in older women who would not be taking OCS. The cause-and-effect relationships in any of these cancers is far from

[3]"Major Advances in the Diagnosis of Skin Cancer" by Lawrence K. Altman, *The New York Times*, January 8, 1980.

clear at present. The designation of the OCS as another cause is, in most cases, only a possibility.

We have, at this time, an unprecedented medical phenomenon. Large numbers of relatively young women are taking a hormone medication that contains two powerful hormonal agents, estrogen and progesterone, in the absence of any defined illness. Currently, teen-agers and even preteen-agers are taking OCS, and quite possibly will be taking them for a greater part of their reproductive lives. We are concerned, and rightly so, because many of us will die of cancer, of one form or another, if present mortality trends continue. We therefore want to examine the data that are available to tell us whether or not the ingestion of OCS can increase our chances of getting cancer.

Case-control studies[4] and cohort studies[5] from Europe and the United States give us the following information about the incidence of various cancers in women who use OCS (World Health Organization, 1978):

1. *Breast neoplasia (benign tumors). When used for two years or more, OCS* decrease *the incidence of this tumor. Increased protection is associated with increased amounts of the progestogen in the medication.*
2. *Breast cancer. No general increased incidence with long term use of OCS. Some subgroups of women, such as those who use OCS prior to the birth of their first child,* may *be at increased risk.*
3. *Endometrial cancer. This form of cancer, rare in women under 40, may be increased in women who use sequential OCS.*
4. *Cervical cancer. Few data are available. A* possible *increased risk with long-term OCS use has been suggested, but not confirmed.*
5. *Ovarian cancer. Insufficient data available.*
6. *Malignant melanoma. No established risk.*
7. *Cancer of the liver. Increased risk with the extended use of preparations that contain higher doses than those commonly used today. Risk increases with age of user.*

ESTROGEN DURING THE MENOPAUSE Estrogen stimulation can produce precancerous changes in the uterus. The development of breast cancer has also been related to the effect of "unopposed estrogen

[4]Cases with the disease (cancer) are matched with one or more control cases selected for certain like qualities such as age and similar use of OCS.

[5]Groups of women using different contraceptives are followed over periods of time.

activity." Yet, without some replacement for the estrogen which our ovaries cease to produce after the menopause, we may be incapable of doing some things that are pleasurable, healthful, or both.

Without estrogen the lining of the vaginal epithelium may thin out. If it does, intercourse can be a physically painful experience. Further advances into the menopausal years can be accompanied by a weakening bone structure that hampers our participation in active sports and may even lead to incapacitating illnesses. Estrogen supplements may predispose us to cancer, yet estrogen may be necessary if we are to retain our capacity to enjoy sex and sports beyond the menopause. What should we do?

Professional groups—physicians, the National Institutes of Health, and the World Health Organization (1979)—have asked the same questions, but for less personal reasons. They have attempted to evaluate the risk-benefit relationship in estrogen supplements for menopausal women, who, by reason of their age, are in high-risk groups for the development of breast and gynecological cancers. Their evaluation was difficult because they needed more information, particularly about:

1. *the natural course of the menopause in the absence of hormone therapy*
2. *alternatives to estrogen*
3. *all aspects of the beneficial and adverse effects of estrogens*
4. *the best way to provide estrogen.*

While waiting for this information to be accumulated, we who are in the perimenopausal phase of our lives need to make our own decisions. These should be informed decisions, based on our knowledge that estrogen not only may increase our risk of having breast and uterine cancer, but that estrogen also benefits our participation in and enjoyment of activities that are important to us. If we decide that, for us, the benefits of estrogen during menopause outweigh the risks, we should increase our surveillance for signs of breast and uterine cancer. Our physicians may perform more frequent breast examinations with possible mammography and endometrial biopsies.

DEALING WITH CANCER

The Importance of Early Diagnosis

The importance of early diagnosis has been stressed. When this diagnosis is positive, and we are told, "Yes, you have cancer," what do we do? Well, first of all, we have a right to be fully informed about the extent of our disease, the proposed diagnostic and therapeutic treatments, and the possible hospitalization that may be involved.

The physician should tell us what kind of cancer we have, how far advanced it is, and what physiological effects it might have on our bodies. The prognosis of our disease with or without treatment should be explained, insofar as it is known.

Questions we might ask concerning proposed confirmatory diagnoses and/or therapeutic treatment include:

1. *What is this procedure called?*
2. *How is it performed?*
3. *What are the chances (percentages) that it will help?*
4. *Will it need to be repeated? If so, how often?*
5. *How long will it take?*

If hospitalization is necessary we will want to know how long we can expect to be away from our families and customary activities so that we can plan accordingly. We need to question the physician about the procedures that will be carried out on us in the hospital—whether they are painful or activity-limiting, or if there are any side effects or complications. What risks are involved in the treatment? If there are alternative treatments for our particular disease, we should ask why the physician has chosen this particular route, and whether it is an accepted procedure or an experimental procedure (McGowan, 1978).

Surgery, radiotherapy, and chemotherapy may be used singly or together. We have seen, for example, in the case of breast cancer, the consequences of the surgery may require considerable psychological, as well as physiological, recovery methods. The same is true for radiation therapy and chemotherapy.

Treatment for Cancer

RADIATION THERAPY Why, we may ask, if radiation can *cause* cancer, is it used to *cure* cancer? We may recall the way in which high-energy radiation affects cells; it works on the genetic material, the DNA. X-rays, and to a lesser extent, ultraviolet rays, can reprogram a normal cell to become a cancer cell.

Radiation therapy or treatment of cancer causes even more profound changes in the genetic apparatus of the cancer cell. The sources of radiation are varied; they may be x-rays, gamma rays from radioactive substances such as cobalt 60, or high-speed electrons from machines such as the betatron. The radiation produced by each has similar effects on the body. It ionizes molecules in the cell. Huge amounts of energy are focused on the rapidly multiplying cancer cells in our bodies. Enough of this energy can destroy all the cancer cells. How does it do this? Probably by causing breaks in the DNA of the chromosomes, and thus interfering with normal cell division. Irradiated cancer cells cannot reproduce normally, and so they die. What about irradiated normal cells? Of course, they may die too, but the skill of the radiation oncologist permits such fine adjustment of the area of treatment (cancer area only) and the dose (amount of radiation delivered to the tissue) that this is unlikely to happen. Cancer cells are slightly more sensitive to radiation than normal cells because they are dividing more rapidly.

When we receive radiation therapy, the cancer in our body will be subjected to ionizing radiation from a large machine at certain time intervals. We and the machine will be alone in a shielded room for periods of time while the ionizing radiation is affecting the biochemistry of our cells. Every effort will be made to irradiate only the cancer cells; however, we may experience side effects related to the area which is treated. For example, treatment of the pelvic area may bring an aftermath of diarrhea. The possible aftereffects of radiation therapy should be explained to us. They will probably not be as uncomfortable as we feared, particularly if the dose adjustment has been carefully planned by the physician in charge.

The outcome of radiation therapy has been generally encouraging. Most radiation therapy centers report that at least 50 percent of their treated patients are still alive, with no evidence of the disease, after five years. Early carcinoma of the cervix has about an 85 percent five-year survival rate with irradiation alone. If we do have cancer, and radiation therapy is recommended as a primary or adjunct treatment, we should be encouraged by these positive results.

CHEMOTHERAPY Like radiation, chemicals can cause cancer and chemicals can cure cancer. The action of the chemicals is more general, however. Radiation therapy is administered to a certain area of our bodies, whereas chemicals travel in the blood to all parts of the body. Effects of radiation therapy are more localized whereas those of chemotherapy are systemic; that is, they affect the entire body (Fig. 8-20).

Chemicals used to treat cancer act as antimetabolites; that is, they disrupt the normal metabolism of the cell. Some inhibit DNA or RNA synthesis. Anticancer drugs may also inhibit protein synthesis; still other drugs inhibit mitosis, which, as we have seen, occurs faster in the cancer cells than in the normal cells. Some cancer-fighting drugs upset the hormonal balance of the body and lead to conditions unsuitable for the survival of cancers. Chemotherapy is based on the premise that cancer cells differ from normal cells in some way that can be exploited by the therapist.

Chemotherapy is used to treat cancers that have spread (me-

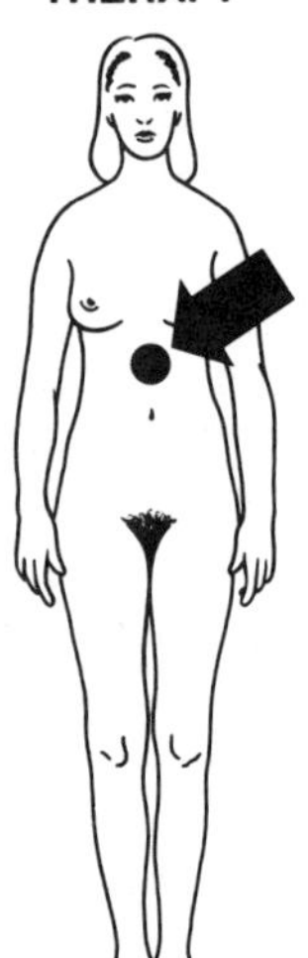

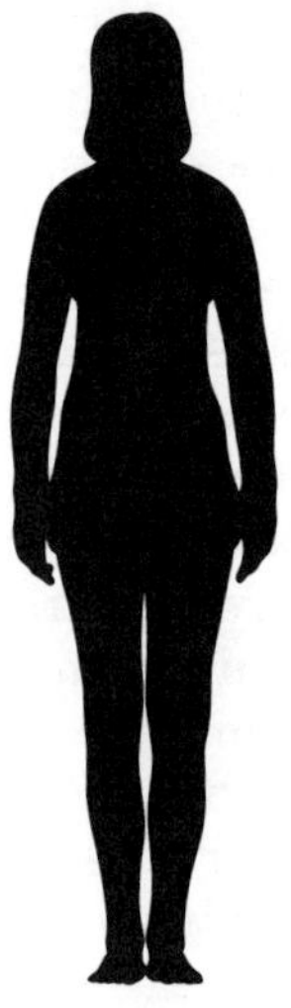

FIGURE 8-20 Effects of radiation therapy versus chemotherapy.

tastasized) throughout the body. Their location cannot be pinpointed, but their growth can be arrested. Drugs are used as adjuvant therapy, for example, when cancer is discovered in several lymph nodes.

Since chemotherapy generally affects tissues composed of rapidly dividing cells, we might expect side effects in these tissues at the same time the cancer cells are being altered. Cells lining the stomach and intestines, the mucous membranes of the respiratory and urogenital systems, the hair follicles, and the bone marrow are all vulnerable to the action of chemotherapeutic agents. So we might experience:

1. *nausea and vomiting from damage to the stomach and intestines*
2. *dryness of the mouth from damage to the mucous membranes*
3. *partial loss of hair from damage to the hair follicles*
4. *infection, bruising, and fatigue from damage to the bone marrow, which produces blood cells.*

The physician will monitor the severity of the side effects through blood counts and other clinical tests. Side effects vary from individual to individual. If a chemotherapeutic agent is prescribed for us, we should know the name of the medication, how it is given (orally, by injection, or intravenously) and how often, what risks are involved (whether the medication is generally used or is experimental), and how it produces its effects. The side effects experienced by others should be reviewed for us. If we know what to expect, we can be participants in our own treatment. Temporary discomforts can be evaluated in relationship to long-term benefits.

Cancer, a disease of cells, can be removed from our bodies when malignant cells are eliminated surgically, destroyed by radiation therapy, or arrested in their growth by chemotherapeutic agents. The ideal anticancer drug is still "out there somewhere" with the ideal contraceptive. No *one* agent has yet been found; in its absence, a variety of different agents are used. Basic research in cell biology may result in improved chemotherapeutic agents that are more sparing of normal, nonmalignant cells.

NEW AND EXPERIMENTAL TREATMENTS New treatments for cancer are being tested. Immunotherapy, a promising new idea in treatment, allies the body's own immune system in the fight against cancer. The importance of the immune system is illustrated by the following facts: (1) Individuals with immunodeficiency diseases have a greater than normal risk of developing cancer, and (2) Individuals treated with immunosuppressant drugs over long periods of time are more susceptible to cancer.

Interferon, a naturally occurring material that inhibits the reproduction of viruses in our bodies, is being examined for its ability to arrest cancer formation. It is possible that interferon slows cancer growth by inhibiting the movement of cancer cells (Brouty-Boyé and Zetter, 1980).

From time to time, we may hear of new drugs, or of drugs offered in other countries and not yet approved for use in the United States. New drugs in our country take from one to four years to get to the general public. This waiting period is necessary so that we will not be exposed unnecessarily to hazardous chemicals. Drugs claimed to have been used "with great success" in other countries are no doubt in the process of being tested by the Federal Drug Administration. It may well be that their toxic effects far outweigh their beneficial effects. It is certainly to be hoped that new treatments for cancer, with demonstrated safety, will be developed in the very near future.

Death from Cancer

Hope that new methods of treating cancer will be found or that somehow a specific cancer will not prove fatal, is something that must never be withheld from the cancer patient even when the prospects of her recovery are very unlikely. No matter how grim the situation, there is always some room for hope. Even now, in the 1980s, when we have so many new advances in cancer detection and therapy, the precise course and outcome of an individual cancer in an individual person cannot be predicted with certainty. Uncertainty leaves room for hope.

If the prospect of terminal cancer comes into our lives, we can always hope, even in hopeless situations, that an arrest or remission of the cancer may occur. Persons who care for patients with cancer that is beyond treatment need to communicate this hope. The ordinary means—including treatments, medicines, and operations—that offer a reasonable hope of benefit without excessive pain, expense, or other inconvenience should be used. Extraordinary means, not obligatory, include treatments that can only be obtained and used at the expense of excessive pain, financial cost, or other inconveniences, and that do not offer *reasonable* hope of benefit. On the other hand, the means we consider extraordinary at this time may be considered ordinary and commonplace within a short period of time. We need to be alert for these changes.

The terminally ill cancer patient would probably prefer to be at home, if this possible. The Hospice concept (Duffy, 1979), a method of cancer patient care more common in England and Ireland than in our country, recognizes in a most particular way the needs of the dying and often makes it possible for the terminally ill patient to die at home if that is her and her family's wish. In the United States, Hospice is not necessarily a separate place such as a hospital ward or a specialized nursing home. It is an integrated approach to caring for the medical, psychological, and spiritual needs of the terminally ill patient when no further curative therapy is indicated. Hospice care of the patient focuses on the patient and her family rather than on the disease and its treatment. At the same time, the physical discomforts of the patient are eased by pharmaceutical and clinical techniques for symptom control, especially pain control. Hospice also attempts to ease the psychological discomfort of the terminally ill through programs to help her and her family as they experience the trauma of progressive disease and the final separation of death (Dunphy, 1976).

Elisabeth Kübler-Ross (1973) tells us that we may achieve peace, our own inner peace, by focusing on and accepting the reality of death. She has, as she tells us, learned from the dying patient that this acceptance has preliminary stages: denial and isolation, anger, bargaining, and depression. We may see these phases as they are lived through in the death of someone close to us. But we will also see the final stage, acceptance.

Those who have the strength and love to sit with a dying patient in the silence that goes beyond words will know that this moment is neither frightening or painful, but a peaceful cessation of the functioning of the body. [Kübler-Ross, 1973]

As women, we are very close to life in the beginning of new life and the nurturing of this life. We have accepted life. We will one day be close to death, our own or that of someone we love. We may then participate directly or indirectly in the preliminaries: denial, anger, bargaining, depression. Acceptance, the final phase, is harder to realize for death than it is for life. We know life, but death is unknown and therefore we fear it. Cancer may bring us very close to death; it may also force us to reevaluate our lives.

REFERENCES

WHAT IS CANCER?

American Cancer Society. *Cancer Facts and Figures.* New York: 1980.

Bayh, Marvella, and Mary Lynn Kotz. *Marvella: A Personal Journey.* New York: Harcourt Brace Jovanovich, 1979.

Braun, Armin C. *The Story of Cancer.* Reading, Mass.: Addison-Wesley, 1977.

Garfinkel, Lawrence, and Edwin Silverberg. *Cancer Statistics, 1979.* New York: American Cancer Society, 1979.

Levitt, Paul M., and Elissa S. Guralnick. *The Cancer Reference Book: Direct and Clear Answers to Everyone's Questions.* New York: Paddington Press, 1979.

Rapaport, Stephen A. *Strike Back at Cancer.* Englewood Cliffs, N.J.: Prentice-Hall, Inc., 1978.

CANCER IN OUR BODIES

Abrahamson, Daniel J., and Paul C. Morton. "Proctosigmoidoscopy in Asymptomatic Men and Women." New York: American Cancer Society, 1978.

Altman, Lawrence K. "Major Advances in the Diagnosis of Skin Cancer," *The New York Times,* January 8, 1980.

Avery, Robert J., Jr., Office of Cancer Communication, National Cancer Institute. Personal communication, 1980.

Barber, Hugh R. K.; Edward A. Graber; and Tae Hae Kwon. *Ovarian Cancer.* Professional Education Publication. New York: American Cancer Society, 1975.

Barnes, Ann B.; Theodore Colton; Jerome Gundersen; Kenneth L. Noller; Barbara C. Tilley; Thomas Strama; Duane E. Townsend; Paul Hatab; and Peter C. O'Brien. "Fertility and the Outcome of Pregnancy in Women Exposed in Utero to Diethylstilbestrol," *New Eng. J. Med.* 320, no. 11 (March 13, 1980): 609–13.

Edmonson, Hugh A.; Brian Henderson; and Barbara Benton. "Liver Cell Adenomas Associated with the Use of Oral Contraceptives," *New Eng. J. Med.* 294 (February 26, 1976): 470–72.

Greenwald, Earl F. "Ovarian Tumors," *Clin. Obstet. Gynecol.* 18, no. 4 (December 1975): 61–85.

Gumport, Stephen L.; Matthew N. Harris; and Alfred Kopf. "Diagnosis and Management of Common Skin Cancers." New York: American Cancer Society, 1974.

Gusberg, S. B., and H. C. Frick. *Corscaden's Gynecologic Cancer,* 4th ed. Baltimore: Williams and Wilkins, 1970.

Herbst, Arthur L.; Howard Ulfelder; and David C. Poskanzer. "Adenocarcinoma of the Vagina," *New Eng. J. Med.* 284, no. 16 (April 22, 1971): 878–81.

Herbst, Arthur L.; Robert Scully; and Stanley J. Robboy. "Vaginal Adenosis and Other Diethylstilbestrol-Related Abnormalities," *Clin. Obstet. Gynecol.* 18, no. 3 (September 1975): 185–94.

Hertz, R., and J. C. Bailer III. "Estrogen-Progestogen Combinations for Contraceptives," *JAMA.* 198, no. 19 (November 1966): 1001.

Lasser, Terese. "Reach to Recovery." New York: American Cancer Society, 1974.

Leffall, LaSalle D., and Maus W. Stearns. "Early Diagnosis of Colorectal Cancer." New York: American Cancer Society, 1974.

Lingeman, Carolyn. "Cancer of the Human Ovary: A Review," *J. Nat. Cancer Inst.* 53, no. 6 (December 1974): 1603–18.

MacMahon, Brian; Philip Cole; and James Brown. "Etiology of Human Breast Cancer: A Review," *J. Nat. Cancer Inst.* 50 (1973): 21–42.

National Institutes of Health. "Estrogen Use and Postmenopausal Women," *Consensus Development Conference Summary* 2, no. 8 (1979).

Reel, Jerry R. "The Mode of Action of Progestogens on Endometrial Carcinoma," in *Steroid Hormone Action and Cancer,* p. 85. New York: Plenum Press, 1976.

Rosenow, Edward C. III, and David T. Carr. "Bronchiogenic Carcinoma" in *Lung Cancer,* Professional Education Publication, pp. 3–15. New York: American Cancer Society, 1979.

Rubin, Philip, ed. *Updated Breast Cancer.* New York: American Cancer Society, 1978.

Rust, Jesse A.; Ivan I. Langley; Edward C. Hill; and Emmet J. Lamb. "Estrogens: Do the Risks Outweigh the Benefits?" *Am. J. Obstet. Gynecol.* 128, no. 4 (June 15, 1977): 431–39.

Sebastian, James A.; Burton O. Leeb; and Richard See. "Cancer of the Cervix—A Sexually Transmitted Disease," *Am. J. Obstet. Gynecol.* 131, no. 6 (July 15, 1978): 620–623.

Strax, Philip. *Early Detection.* New York: Harper & Row, 1974.

———. "The Role of Thermography as Compared with Mammography," in *Updated Breast Cancer,* p. 14. New York: American Cancer Society, 1978.

Ulfelder, Howard. "The Stilbestrol-Adenosis-Carcinoma Syndrome," *Cancer* 38 (1976): 426–31.

Wallace, Robert B.; Barry M. Sherman; Judy A. Bean; James B. Leeper; and Alan E. Treloar. "Menstrual Cycle Patterns and Breast Cancer Risk Factors," *Cancer Research* 38 (November 1978): 4021–24.

World Health Organization. "Steroid Contraception and the Risk of Neoplasia," Technical Report Series, no. 614. Geneva: World Health Organization, 1978.

Wynder, Ernest L.; Hideaki Dodo; and Hugh R. K. Barber. "Epidemiology of Cancer of the Ovary," *Cancer* 23 (February 1969): 352–70.

DEALING WITH CANCER

Abrahamson, Norman. "Radiation Therapy: What Is It?" *The Southern Med. J.* 67, no. 11 (1974): 1333–36.

Brouty-Boyé, Danièle, and Bruce R. Zetter. "Inhibition of Cell Motility by Interferon," *Science* 208 (May 2, 1980): 516–18.

Duffy, Ben. "Hospice: The Secret of Care Is in Caring." *Physician East,* June 1979, pp. 14–16.

Dunphy, J. Englebert. "On Caring for the Patient with Cancer," *Bull. Am. Coll. Surg.,* October 1976, pp. 7–14.

Krakoff, Irwin A. "Cancer, Chemotherapeutic Drugs." New York: American Cancer Society, 1977.

Kübler-Ross, Elisabeth. *On Death and Dying,* New York: Macmillan, 1973.

McGowan, Larry, ed. *Gynecologic Oncology,* New York: Appleton Century Crofts, 1978.

9
Venereal Disease in Women

The spread of venereal disease in the country is beyond control, particularly among teen-agers. As a research scientist at the Center for Disease Control in Atlanta puts it, "Insofar as VD is concerned, the genie is out of the bottle." He tells of his interior decorator who went on a week's vacation with her husband. They left their sons, ages 13 and 14, unsupervised at home. When they returned both boys had gonorrhea.

The statistics on the incidence of communicable diseases in this country (Fig. 9-1) may shock us. Gonorrhea leads all the rest, and the figure of over 1 million cases per year may represent only half the actual cases because of underreporting. Most gonorrhea cases are treated by private physicians who may or may not report them. Syphilis follows chicken pox as the third most common communicable disease. The figures for syphilis are probably accurate. Untreated syphilis has such drastic consequences that most cases are reported as they are diagnosed.

We live in a country that prides itself on its scientific and technological advances, and yet VD, which decimated populations in Europe hundreds of years ago, is rampant among us, particularly among young people. Venereal disease, for all practical purposes, means sexually transmitted disease. We can only get venereal disease from sexual contact with an infected person. The patterns of our sexual contacts change during our lives. Each of us has her own value judgments by which she regulates her relationships, sexual and otherwise. The reality of venereal disease is that it occurs in a variety of moral contexts. We cannot substitute a moral judgment for a medical cure; however, we can become more aware of factors that increase the likelihood of getting VD.

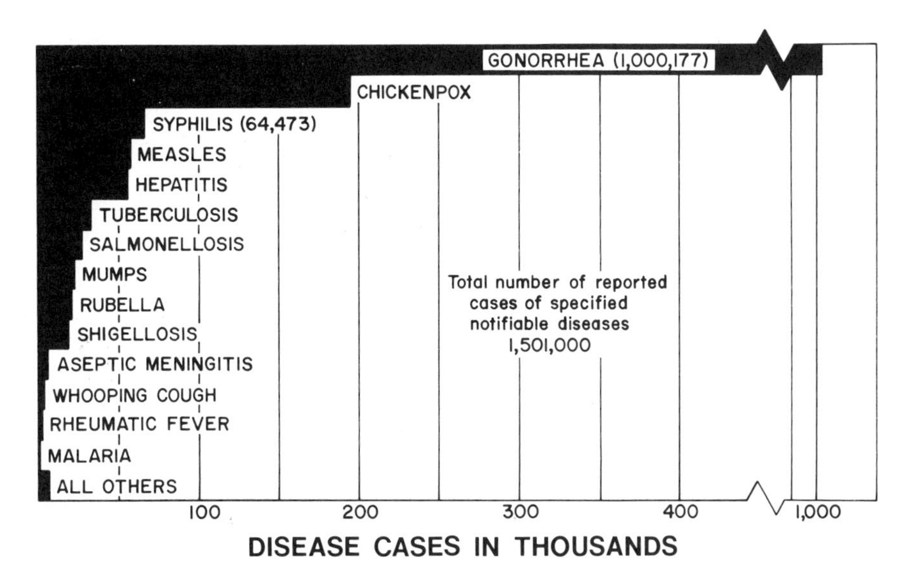

FIGURE 9-1 Communicable diseases: Number of reported cases in the United States in 1977. From the Center for Disease Control, Atlanta, Georgia.

Casual sex with many partners is the pattern of sexual activity for many women who get VD. This may or may not be true in every case, since sex requires two persons and we may not know our partner/lover as well as we think we do. Although venereal diseases can and do result from rape, most venereal diseases develop in our bodies as a direct result of freely chosen sexual activity. Since this is true, we need to ask ourselves some basic questions: "How well do I know my sexual partner?" "Are my partner's only sexual contacts with me?" If we don't know our partner well, and if he has other contacts we may be at risk for getting VD.

Dr. John W. Grover, a physician who copes with the consequences of venereal diseases in young people, believes that "Venereal diseases are cruel infections to have and to transmit. The factors that lead human beings, especially young people, to become sexually active are . . . complicated and poorly understood." His interviews and case studies reveal some prevalent attitudes and practices in young Americans:

Case 1: *Purvis is a sophomore at a nearby college, and I met him while on a TV panel discussing sexuality. He defines promiscuity as "more than one girl a night." He feels that sex is better in "meaningful relationships" but it is easier for boys to move from one "meaningful relationship" to another than it is for girls. He's a youth counselor in his hometown church.*

Case 2: *Charlene is 19 years old, and has been living with Johnny for four months—or, rather, he with her, since he moved into her apartment. She has had gonorrhea several times in the past, but is ready to settle down now. She doesn't believe he's ready yet, so she's willing for him to see and have intercourse with other women until he feels ready to live with her permanently. She says he's only been with 20 or 25 girls and still isn't experienced enough. On direct questioning, she herself admits to 50 or 60 different men, and believes this entitles her to settle down.*

Case 3: *Cheryl is the daughter of a college friend. She had one legal abortion at 18, none since because of birth control pills. When I asked her if she was in a stable relationship now, she said, "Well, more or less. Only one or two guys." When I asked her if she had ever engaged in intercourse on first dates, she said, "Not so much anymore, and then only if I like him."*

Case 4: *Emily is 12 years old. A few months ago, she was admitted to the hospital with lower abdominal pain, possibly due to appendicitis. However, her signs and symptoms were atypical, and the doctor ordered numerous laboratory studies to confirm the diagnosis. A bacterial culture grown from her vaginal secretions gave the answer: a pelvic infection with peritonitis due to gonorrhea. Her boyfriend was 13.*[1]

Attitudes toward sex, expressed by actions of these young people, are formed by complex interrelationships among conditions and persons in their environment. Some teen-age sexual behavior is the result of impulse rather than attitude. Parents and teachers may have very little input into the formation of these attitudes. For the present, we can only deal with the possible consequences of these sexual attitudes and practices, which may include venereal disease.

If we know what the venereal or sexually transmitted diseases are, their symptoms, and where to go for help, we may be able to escape the permanent consequences of these "cruel infections." These consequences may be more severe in women, and may include recurrent infection, permanent sterility, and lasting damage to unborn children.

THE NATURE OF COMMUNICABLE DISEASES AND THE BODY'S DEFENSES

Disease in our bodies may result from the interaction of a foreign invader with our body's defense mechanisms. Foreign invaders may include bacteria, viruses, protozoans (one-celled animals), yeasts, and

[1]From the book *VD: The ABC's* by Grover and Grace. © 1971 by John Wagner Grover, M.D. and Dick Grace. Published by Prentice-Hall, Inc., Englewood Cliffs, New Jersey 07632.

sometimes insects. Because most of these organisms are very small, they are generally referred to as microorganisms or microbes. When these microbes are introduced into our bodies, they may become part of an infectious disease cycle which includes:

1. *the agent—the particular microbe which may cause the disease*
2. *the reservoir—any infected person, with or without symptoms*
3. *the place of exit—where the agent leaves the reservoir*
4. *the method of transmission—how the agent leaves the reservoir*
5. *the place of entry—where the agent enters the newly infected individual*
6. *a susceptible host—any person or object in which the agent can grow and reproduce.*

The general infectious disease cycle for sexually transmissible diseases (STD) is outlined in Table 9-1.

Our body's defenses against an infectious agent include (1) the intact skin and lining of the internal organs, (2) a general inflammatory response, and (3) a specific immune response. If these defenses fail, we get the disease. Each individual has unique strengths and weaknesses in her defense mechanisms. That is why some people who are exposed to a disease agent get the disease while others do not.

When microbes gain entry through a break in the protective surface, they are dealt with, in a general way, by an inflammatory response. The pain, reddening, and swelling of a cut or sore are consequences of the inflammatory response and show that our body's defense mechanisms are working. The reddening characteristic of the inflammatory response is caused by the dilation of blood vessels at the site of the injury. Swelling occurs because fluid filters into the tissues. The pain that accompanies redness and swelling is a result of the distension and the effect of released substances on nerve endings. During the inflammatory response certain types of large white blood cells, called neutrophils, surround the microbes. Figure 9-2 shows a neutrophil engulfing some microorganisms that cause gonorrhea. The neutrophils interact with the microbes and make them powerless to harm us. The pus we see at the site of an injury represents the product of the microbe-neutrophil interaction.

TABLE 9–1 Infectious Disease Cycle for STD

Agents	Bacteria, viruses, protozoans, yeasts, insects
Reservoir	Infected human beings
Place of Exit/Entry	Penis, vagina, mouth, anus
Method of transmission	Sexual contact
Susceptible host	Another human being

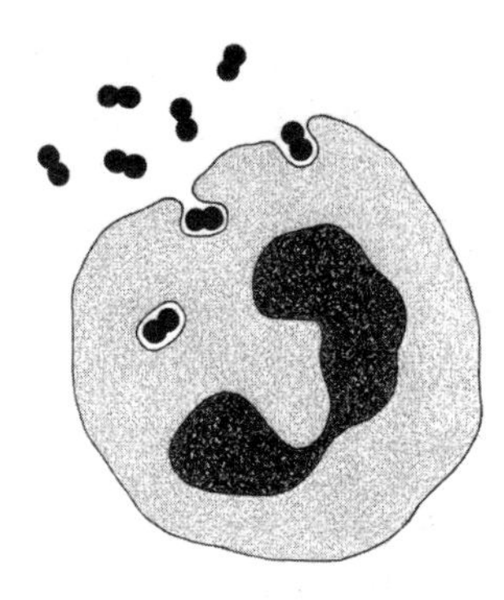

FIGURE 9-2 Specialized leucocyte (neutrophil) ingesting gonococci.

The more specific defense mechanisms involve smaller white blood cells, called lymphocytes, which can accommodate their response to a particular invading organism. The infectious agent acts as an antigen and the B-lymphocytes produce antibodies in response to them.

Immunity

Because of a previous production of antibodies by the lymphocytes we may have a specific, active immunity toward an infectious agent. This specific immunity can be a result of our having had the disease, or having had a vaccination for the disease. When we are vaccinated, a small amount of the agent is injected into our bodies. Sometimes the agent has been altered to decrease the chances of its being harmful. Our specific defense mechanisms respond by producing antibodies. If we are reexposed at some time in the future the lymphocytes that formed the first set of antibodies remember how they did it and quickly produce a sufficient number of antibodies to protect us from getting the disease.

In addition to the active immunity described above, we may have a passive immunity to a disease by being injected directly with the antibodies. These antibodies are generally produced in experimental animals. Passive immunity is also conferred on the fetus during intrauterine life when antibodies produced by the mother pass through the placenta. Passive immunity is short-lived. It is used when fast protection is needed for a person who has been or will be exposed to the antigen. Why don't we have shots, injections of antibodies, to protect us against venereal diseases? Part of the answer to that question lies in the fact that these microbes only live in the human host. It is therefore difficult to produce antisera, containing the antibodies, in large quantities.

Research microbiologists are working to develop vaccine-like substances to protect us against venereal diseases in the same way we are protected against other communicable diseases. We cannot count on this kind of protection at the present moment. Even the fact that we have had a sexually transmitted disease will not guarantee our protection at the next exposure.

Antibiotics

Antibiotics such as penicillin and tetracycline are our chief weapons against some venereal diseases. These antibiotic substances have been remarkable allies in curing sexually transmitted diseases such as syphilis and gonorrhea.

Antibiosis, from which we get the name "antibiotics," refers to the unbloody battle among living things for the resources necessary for life. These resources are limited. In order to win the battle and to survive, some microorganisms that live in the soil have developed the ability to produce chemicals that are harmless to themselves but poisonous to competitors. One of these soil microbes, *Penicillium notatum,* produces such a chemical. This chemical is called penicillin. Penicillin is an antibiotic against a wide range of competitors including the bacteria causing syphilis and gonorrhea.

The development of different antibiotics and their use in fighting infectious diseases has been one of the notable achievements of modern medicine. When defense mechanisms in our bodies do not combat microorganisms effectively, antibiotics may be indispensable. However, a disease must be recognized and diagnosed correctly for a prescription of antibiotics to be useful. In addition to this, some microbes develop penicillin resistance. While we might prefer to believe that antibiotics are the solution to all the problems caused by sexually transmitted diseases, the facts indicate otherwise.

SYMPTOMS, DIAGNOSES, TREATMENTS, AND AFTEREFFECTS OF VENEREAL DISEASES

Gonorrhea

Gonorrhea is the number one communicable disease in the United States today (Fig. 9-1). Gonorrhea has been classified by the Center for Disease Control as pandemic, meaning that it occurs in persons who live in disparate communities. It is not isolated in a single region of the country, nor is it characteristic of a given social class. The agent of the disease is a bacterium, *Neisseria gonorrhoeae,* commonly called the gonococcus. It can only grow and survive in human tissues. Gonorrhea is transmitted to adults only through sexual contact with an infected person.

Most men who have the disease manifest symptoms, including a penile discharge that usually contains pus. On the other hand,

60 to 80 percent of women having the disease will show no external signs. In the absence of gonorrhea symptoms, what do we do if we suspect that we may have the disease? First of all, we should know our partner. He may notice the discharge. Hopefully, he will tell us about it.

If a woman does develop symptoms, they appear two to nine days after exposure. Symptoms include painful urination, a yellowish discharge, and vaginal pain or pelvic discomfort. Anal and pharyngeal infection with gonorrhea may be a result of genital-anal or oro-genital sexual practices. Rectal pain or a sore throat will be symptomatic in these cases.

The gonococcal microorganism multiplies in the cervix of the uterus (Fig. 9-3). If they are unopposed by treatment, the gonorrhea infective agents travel to the oviduct where they continue to multiply. While the woman who has the disease is generally unaware of it, the pus-like material containing the microbes spreads to the ovary and to the other pelvic organs where it produces an inflammatory response.

Systemic infection with the gonococcus occurs in 1 to 3 percent of persons who harbor the microorganism. Arthritis is an outcome of disseminated infection with the gonorrhea microorganism. It is one of the most common causes of acute arthritis in young adults in the United States (Handsfield, 1975).

Complications resulting from untreated gonorrhea infection in women can include inflammation of the oviduct, lower abdominal pain, especially during intercourse, and irregular menstrual cycles with prolonged bleeding. If the symptoms which do appear in the latter stages of the infection are ignored and the infection is not treated the complications are serious. Permanent sterility may be the result of an untreated gonorrhea infection. If both oviducts become infected, or they close from permanent scarring, they cannot

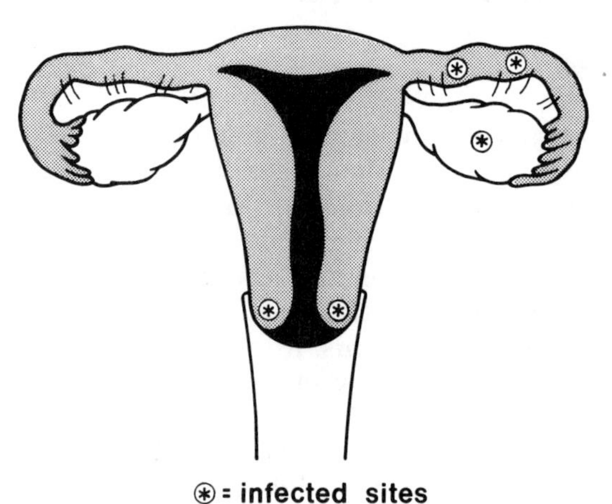

FIGURE 9-3 Growth of gonococcus in reproductive tract of infected female.

transport the fertilized ovum to the uterus. Some authorities estimate that as many as 100,000 young women become sterile each year as a result of gonorrhea infections.

Oral contraceptive steroids taken to produce temporary sterility may actually lead to the permanent sterility induced by gonorrhea because (1) they may increase the number of sexual contacts when fear of pregnancy is not a deterrent, and (2) they alter the environment of the vagina to favor the survival of the gonococcus. Whereas perhaps 40 percent of women not on "the Pill" may get gonorrhea when exposed to it, more than 90 percent of those who take OCS will probably contract gonorrhea if they are exposed to it (Fiumara, 1972). Women who have had untreated gonorrhea for a long time are not only sterile, but they lapse into a state of chronic invalidism with numerous symptoms of general ill health. These women seem to become depressed and generally bitter about life. No wonder!

Symptomatic, asymptomatic, or suspicious, a woman who believes she may have been exposed to gonorrhea should seek help at once. The examining physician should perform a complete pelvic examination and take a sample of material from the cervix, urethra, anus, and pharnyx. This sample will be analyzed for the presence of gonococcal organisms.[2] If the analyses indicate gonorrhea infection, treatment is prescribed. In the case of a positive diagnosis of gonorrhea, we should inform our sexual partners. Obviously, if we all did this, the pandemic proportions of gonorrhea could be brought under control.

Before accepting any form of treatment for gonorrhea, or for any other disease, we should be informed of the doctor's diagnosis and the findings upon which the diagnosis is based. We should ask the name of the drug we are getting and the dosage which is prescribed, and we should make a written record of both diagnosis and treatment.

The recommended treatment for gonorrhea is an injection of procaine penicillin G. The standard dose is 4.8 million units, divided in half. One half of the dose is injected into the large muscles in each of our buttocks. Another drug, probenicid, should be given at the same time. Its function is to increase the effectiveness of the penicillin. The dosage is 1 gram, and it is given by mouth. If someone is allergic to penicillin or to probenecid the alternative treatments

[2]According to Public Health Citizens Research Group, three products used to detect gonorrhea in women are inaccurate and should not be marketed. One of the products, Gonopat, produced by Becton, Dickinson and Co., is not being marketed as yet. The others are Gonosticon, Dri-Dot (Organon, Inc.), and FGT-H (Fisher Scientific Co.). Both companies say that their products are intended to be used only for screening purposes (*Wall Street Journal,* June 6, 1980, p. 3). If we have positive gonorrhea tests, we should be certain that test results are based on routine culture methods as well as preliminary screening tests.

are to give tetracyclines by mouth for five days or to give a single intramuscular injection of spectinomycin. Another alternate treatment for someone who is not allergic to penicillin is to give 3.5 grams of amphicillin or amoxicillin along with 1 gram of probenecid. All of these are given by mouth. Regardless of the mode of treatment chosen, the goal is the same: to introduce a large amount of antibiotic into the bloodstream for a short period of time.

The penicillin injection does not conclude our treatment. Antibiotics may cure only 90 to 95 percent of cases on the first try and therefore we should return for a checkup in a week, and for two more checkups during the following week. In the meantime, we should refrain from any kind of sexual activity, or, at least, insist that our male partner wear a condom.

In general, if we are sexually active with a variety of partners, we are well advised to have a pelvic examination and a cervical smear taken three or four times a year. The early diagnosis of gonorrhea in the absence of any symptoms can save us from later symptoms, including sterility. Informing our contacts if this diagnosis is positive can also help reduce the incidence of the disease.

Other Inflammatory Diseases

Nonspecific vaginitis (NSV) and nongonococcal urethritis (NGU) represent another category of sexually transmitted inflammatory disease. A variety of microorganisms cause NSV-NGU. Men who have the disease may have a discharge from the penis. Most women who have the disease are asymptomatic. They may have inflammatory responses in the pelvic area or vaginitis after sexual activity with an infected person. Among the agents that cause NGU-NSV in women are Chlamydia, implicated in about 50 percent of cases, and T-mycoplasmas, implicated in about 30 percent of cases. Chlamydial infections may be a major cause of cervicitis. It is not known at the present time whether scarring of the oviducts, which is known to result from untreated gonorrhea, can be caused by Chlamydial infection, as well. Epidemiological studies in Europe have shown that women who have had Chlamydial infections have a lower fertility rate than that of the general population. Tetracycline is generally used to treat Chlamydial infections. At this time no systematic studies have been directed toward determining the effectiveness of this form of antibiotic therapy.

Not all vaginitis is "nonspecific." Some cases are known to be caused by a protozoan, *Trichomonas vaginalis,* or by a yeast, *Candida albicans* (Monilia). The diagnosis and treatment of trichomoniasis is fairly straightforward. About one out of every five women has trichomoniasis during her sexually active life. Although trichomoniasis is usually transmitted during sexual activity, we can get it from damp washcloths, towels, or swimming suits used by infected persons. Symptoms of infection occur 4 to 48 hours after exposure. They include frequent and painful urination, inflammation of the vulva, and, possibly, severe abdominal pain. A frothy, green-grey, foul-smelling vaginal discharge is characteristic of the disease. The trichomonads can be seen under a microscope if some of the discharge is put on a slide. A Pap smear may also reveal the organism.

Trichomoniasis is effectively treated by a compound called metronidazole (tradenamed Flagyl). The male partner is treated with an oral preparation of this compound. Women may use a vaginal suppository. Flagyl has many side effects and should not be taken by anyone who has a history of blood diseases or central nervous system disorders. It kills white blood cells, and can cause nausea, diarrhea, cramps, or dizziness. It can cause violent nausea if alcoholic beverages are drunk during treatment, although a single glass of beer or wine per day is considered safe.

Monilia *(Candida albicans)* is a yeast-like organism that is normally present in the vagina of about 20 percent of nonpregnant women. The population of Monilia may be increased when antibiotics kill other vaginal flora. Diabetes, pregnancy, and the use of oral contraceptive steroids may also increase the Monilia growth and lead to vaginitis.

During infection with Monilia the vulva may become red and swollen. A common symptom of Monilia infection is a vaginal discharge, which may be thick and white (like cottage cheese) or thin and watery. Many women have no symptoms when they have the disease, but sexual partners can develop sores or inflammation of the penis.

Positive confirmation of Monilia infection can be made by finding the organism in the vaginal discharge or in a vaginal smear. A vaginal cream or suppository containing either nystatin or miconazole is prescribed.

Neither trichomoniasis nor Monilia infections usually have lasting complications. They are more annoying than anything else, and may give us a feeling of being unclean. A trichomonad infection, usually symptomatic, may accompany a gonorrheal infection, which is usually asymptomatic. If we have a trichomonad infection we should ask for an examination for gonorrhea.

Genital Herpes (Herpes Virus Type II, Herpes Progenitalis)

After a three-to-five-day incubation period, sexual contact with an infected partner may give rise to symptoms that include sores rather than discharges. Clusters of tiny fluid-filled blisters may appear on the labia, where they can be seen as well as felt, or in the vagina, where they are not so obvious. At the same time, women who have this sexually transmitted disease may experience symptoms that include swollen lymph glands, aching muscles, and fever.

Herpes is a disease caused by a virus and so far it is a disease without a cure. Physicians in private practice see more cases of genital herpes than they do of gonorrhea and it is second (after gonorrhea) among the most frequently treated diseases in clinics.

At the present time some of the symptoms of genital herpes can be treated, but there is no sure cure for it. Like a cold sore, which is caused by a related virus, the symptoms of genital herpes can recur during times of stress, or when the defense mechanisms of our bodies are weak. The bleeding phase of the menstrual cycle also seems to be a time when women experience recurrent symptoms.

The possible link between herpes virus II and cervical cancer has not been confirmed; nevertheless, it has been recommended that women who have regular intercourse with a man who has genital herpes should have a Pap smear every six months. The Pap smear will show the presence of the virus if it is there, and it will also indicate precancerous conditions in the cervical cells.

Genital herpes is more common in women than in men and it can also be more serious. During normal childbirth, the infant passing through the vaginal tract of an infected mother may be infected with the virus. For this reason, Cesarean section deliveries for herpes-infected women have risen dramatically in the past few years.

Although treatment for genital herpes is not yet available, the symptoms, such as sores, can be treated with warm baths and the avoidance of tight clothing, particularly pantyhose. A self-help group for persons who have genital herpes has been formed and those who wish to have some assistance in coping with this disease may write to HELP, P.O. Box 100, Palo Alto, Calif., 94302.

Syphilis

Syphilis is epidemic rather than pandemic; that is, it occurs in communities but seldom on the nationwide scale of other sexually transmitted diseases. Unlike any of the other venereal diseases, untreated syphilis can do more than cause debility and discomfort. It can cause death.

Syphilis has afflicted many famous and infamous individuals throughout history. The agent of this disease is a highly mobile corkscrew-shaped bacterium called *Treponema pallidum.* Within seconds after contact the bacterium can be transmitted from an infected person, penetrating the intact skin and reaching the bloodstream of a susceptible host.

The stages of syphilis are described according to the symptoms that occur in each stage. The time elapsed between infection and the progressive stages of the disease varies from one person to another and may reflect variability in the individual's defense mechanisms. The usual incubation period is about six weeks.

The primary stage of syphilis is evident from one to six (usually six) weeks after infection. It is characterized by painless sores called chancres, which appear at the point of sexual contact with the infected person. These sores may appear on the cervix, vaginal walls, labia, clitoris, urinary meatus, lips, or anus. Because chancre sores go away even if the disease is not treated, the individual may think that she is cured. The disease progresses however, and symptoms of secondary syphilis usually appear two to ten weeks later.

During the secondary stage of syphilis the infected person may have a skin rash in the form of raised bumps all over the body. Other symptoms include loss of hair, swollen joints, swollen lymph nodes, and general malaise, including loss of appetite, nausea, constipation, and fever. As these symptoms will motivate most persons to seek some kind of medical assistance, most cases of syphilis in this country seldom advance beyond the secondary stage.

Two-thirds of untreated syphilitics recover. They show no symptoms and have developed immunity toward the recurrences. This is one of the few instances of immunity from VD. However, this immunity comes only at a great risk. The untreated person stands a chance of being among the one-third who enter the tertiary stage of syphilis. In this stage, victims develop cardiovascular, visceral, and nervous disorders that are eventually fatal.

If we suspect that we might have syphilis, we should seek immediate diagnosis. During the primary stage diagnosis can be made by the direct observation of the spirochete in the material from the sore. Later, during the secondary stage, the syphilis patient will have formed antibodies that can be detected in the blood. Several blood tests are available that indicate the presence of the antibody. Most states require one of these blood tests for syphilis before marriage and childbirth.

If syphilis is diagnosed, treatment involves the injection of over 2 million units of penicillin into the bloodstream on the first day of treatment. Two more injections are given three days apart. A reexamination should be performed one month after treatment, during which time no sexual activity should be undertaken. Further reexaminations are recommended every three months for a year after the initial treatment.

Congenital syphilis in the newborn, which may be an effect of maternal syphilis, will be examined below.

Other Venereal Diseases

Insects can cause venereal diseases. Warts on the genital area are spread venereally. Some venereal diseases are rare in this country

TABLE 9-2 Venereal Disease and Sexually Transmitted Diseases: Summary Chart

Disease	*Agent (Type)*	*Symptoms in Women*	*Treatment*	*Complication(s)*
Gonorrhea	Bacterium	60–80%, no symptoms; pain when urinating; vaginal discharge; PID[a]	Penicillin	Sterility; disease in newborn
Trichomoniasis	Protozoan	Frothy vaginal discharge; vaginitis	Metronidazole (Flagyl)	Rare
Candida (Monilia)	Fungus	Vaginal discharge; vaginitis	Nystatin (vaginal suppository)	None
T-mycoplasma	Bacterium	Vaginitis; PID	Tetracycline (?)	
Chlamydia	Bacterium	Vaginitis; PID	Tetracycline	Eye disease of newborn
Herpes (genital)	Virus	Cluster of painful blisters on genitals	None available	Herpes infection of newborn
Syphilis	Bacterium	(1) Chancres (2) Rash	Penicillin	Late syphilis; congenital syphilis
Genital warts	Virus	Warts in genital area	Cut or burn off	None
Pubic lice	Insect	Itching sore in genital area	Benzyl benzoate; Y-benzene; hexachloride	Impetigo; eczema
Scabies	Insect	Burrows on skin	Benzyl benzoate; Y-benzene; hexachloride	Impetigo; eczema

[a]PID stands for pelvic inflammatory disease.
Pelvic infections may be caused by (1) intrauterine trauma such as D and C, or postpartum and postabortion events; (2) postoperative events; and (3) bacteria (Eschenbach and Holmes, 1975).

Source: U.S. Department of Health, Education and Welfare, "Sexually Transmitted Disease Summary," Pub. no. 00-3380 (Atlanta: Public Health Service, Center for Disease Control, 1980).

and rare in women. The principal venereal diseases we may contact, either directly or indirectly, are listed on Table 9-2, which has been adapted from the Sexually Transmitted Disease Summary prepared at the Center for Disease Control, Atlanta.

Venereal Diseases that Affect the Newborn

The woman with venereal disease runs a high risk of passing her infection on to her baby. Some forms of VD are transmitted to the fetus prior to birth while others are transmitted during parturition (Remington and Klein, 1976).

Herpes Simplex Virus. Herpes simplex (HV-2) may be acquired from the maternal vagina during parturition. If the child is delivered normally, a 50 percent risk of infection is estimated. For this reason women who have active herpes infections may have Cesarean section deliveries. It is necessary to perform the Cesarean section before the membranes rupture. A child infected with congenital herpes may have encephalitis, which is often fatal. Skin eruptions, eye inflammations, and other symptoms similar to adult herpes are often seen in the infected newborn.

Syphilis. The premarital blood test for syphilis has helped lower the incidence of congenital syphilis. Every woman who becomes pregnant should be tested for syphilis in the first trimester of her pregnancy. Unfortunately, some women do not enter the health care system until the time of delivery.

The outcome of pregnancy when the mother has syphilis depends on the stage of development of her syphilitic condition. In women who have primary or secondary syphilis, about 50 percent of children may be born with congenital syphilis. If the mother is in the last stage of syphilis, with antibody production, the child has a 70 percent chance of being born *without* congenital syphilis.

Gonorrhea. Passage through the birth canal of a gonorrhea-infected mother can transmit the disease to the newborn. Arthritis and eye infections are the most common symptoms of the disease in the newborn. In most states it is mandatory to use a 1 percent silver nitrate solution to treat the eyes of the newborn. This will destroy the microorganisms that produce the eye infection.

Candida Albicans. Infections with *Candida albicans* can be serious in the newborn. If the infection spreads throughout the system, death or mental retardation can follow. Lesions on the mouth or skin are common localized expressions of the infection.

WHAT CAN WE DO ABOUT VD?

Venereal diseases are social diseases, spread by human sexual contact. If we are responsible persons we have obligations to ourselves and obligations toward others in all the areas of our lives, including sexual relationships. Our failure to fulfill these obligations will contribute to the already rampant spread of the venereal diseases in America today, where teen-agers are the most likely group to be affected.

Obligations to Ourselves. Good hygiene is a good start. Frequent showers or baths, using our own dry towels and washcloths, as well as general cleanliness of the genital areas before and after sex are simple and cheap precautions. We also owe it to ourselves to know our contacts and to be selective, because the superpromiscuous are often the superinfected.

Regular medical checkups, more frequent if we have several sex partners, are also a good form of insurance against venereal diseases and their lasting effects. At the time of the examination, we need to be sure that a pelvic exam and a vaginal examination with a speculum are done (Figs. 9-4 and 9-5). If we do have gonorrhea or syphilis and are treated for it, we need the follow-up checks to be certain that the antibiotics have done their work.

Obligations to Others. The most elementary sense of responsibility for others would dictate that we inform our sexual partners if we have a positive diagnosis of a sexually transmitted disease. Even if we have never had a venereal disease, we should be willing to participate in education programs at whatever level we find it possible to make a contribution. The factual content about sexually transmitted diseases is simple; the challenge is in the *way* this material is presented. Negative and unrealistic attitudes about venereal diseases are hard to overcome.

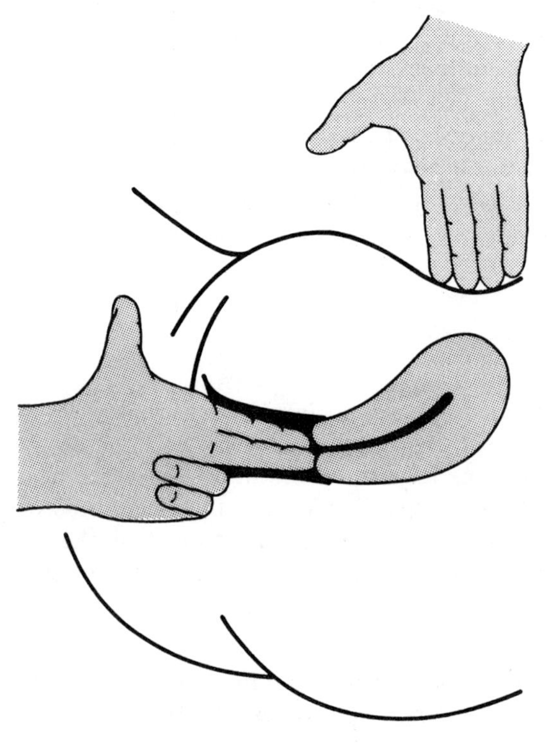

FIGURE 9-4 Pelvic examination.

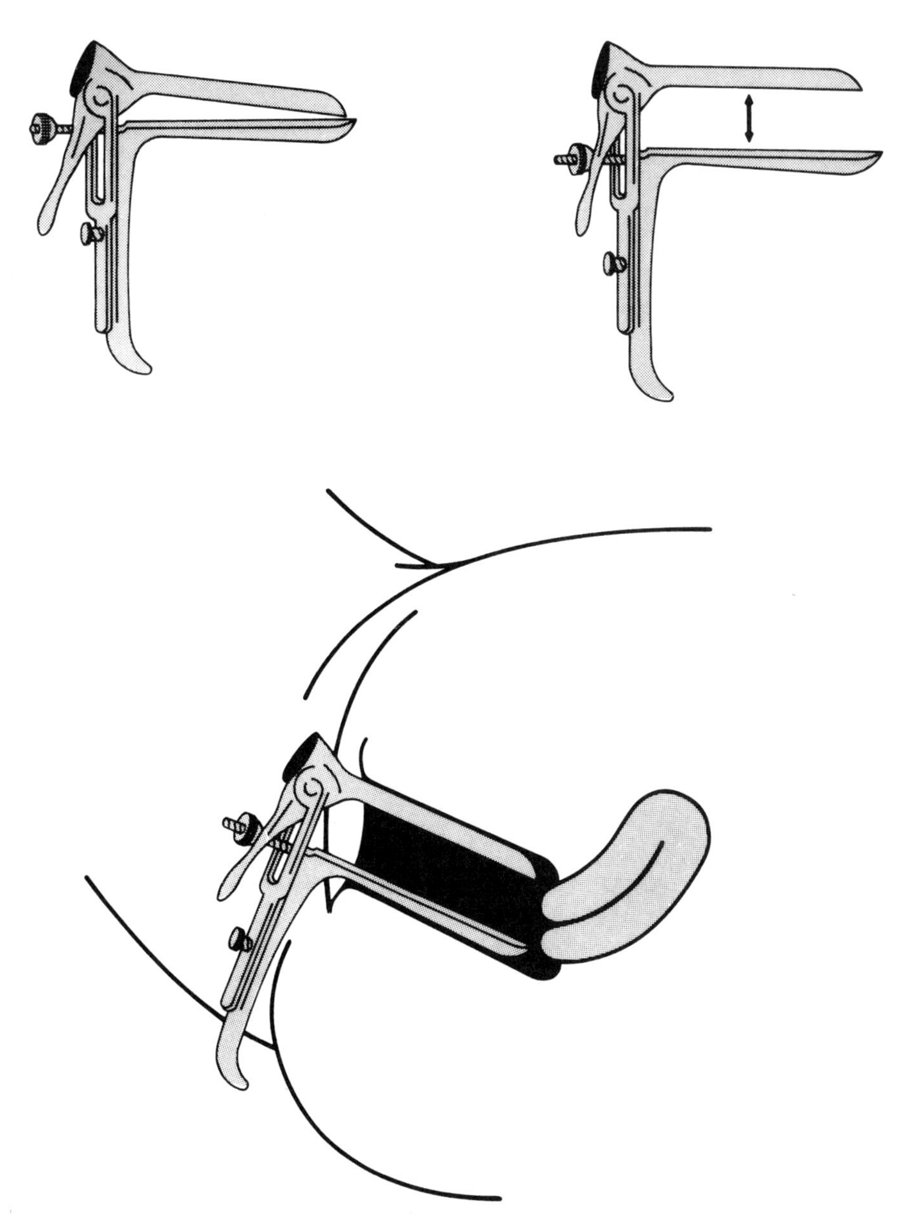

FIGURE 9-5 Vaginal and cervical examination with speculum.

The Condom: Why Not?

The condom is used as a contraceptive more frequently than we might expect. (See Chapters 4 and 6.) Condoms are cheap, easily available, and, if they are used correctly, protect their users against venereal disease as well as pregnancy. Sexually active young teenagers, who frequently do not plan their sexual activity or their protection against the unwanted consequences, could benefit greatly from the use of condoms.

The condom has many advantages, especially for teen-agers (Felman, 1979). It is convenient to keep handy in purse or pocket and it is relatively cheap. Teen-agers who may have different sex partners can know they are being protected against pregnancy and VD if the condom is used. The male, who is generally the initiator of sexual activity, takes the responsibility for protecting his partner. A woman who is sexually active can carry condoms in her purse to protect herself against VD in doubtful situations, even if she is using OCS or is wearing an IUD.

Given these advantages, it is surprising that sexually active teen-agers are not encouraged and advised to use condoms more frequently. Parents and educators need to face realistically the facts of teen-age sexual activity (see Chapter 3), and to be prepared for the consequences if simple protective measures like the condom are ignored.

Medical Help

A woman who suspects she has a venereal disease has two main sources of help: private and public. She may go to her own gynecologist or she may seek the anonymity of a venereal disease clinic. Venereal disease clinics throughout the country vary widely in the quality of their personnel and services; in general, the better clinics are located in large cities and are associated with large medical centers.

Some local communities have telephone answering services that give free health information, including VD facts. Operation Venus (1-800-523-1885) is a nationwide 24-hour answering service that gives free health information to anyone concerned about VD in the presence or absence of symptoms.

National Defense against VD

Education and research are two federally funded attempts to help us deal with venereal diseases. In a continuing education program directed toward medical personnel, the government has attempted to establish a core curriculum of VD facts, diagnoses, and treatments. The training of nurse practitioners for VD clinic work has been a successful outcome of this program. Nurses are trained to do most of the routine work of the clinic, including the pelvic exam (Thompson, 1980).

Efforts such as these should greatly enhance the efficiency of our national health care system. Unfortunately, even *today,* in this country, a 16-year-old girl can have a venereal disease which no one finds until it is too late. A total hysterectomy may save her life, but it obviously ends her reproductive life.

Research efforts are being directed toward the development of vaccines against sexually transmitted diseases (Wong et al., 1979) and toward the screening of antibiotics for the treatment of these diseases. Antibiotics may become useless to treat a specific disease if the microorganism develops a resistance to the antibiotic. The nature of penicillin resistance is being investigated in an academic way (Perine et al., 1977) for practical applications to the problem of controlling the spread of venereal diseases. Unfortunately, research and education efforts seem to have little effect on the high incidence of venereal diseases in our sexually active population.

REFERENCES

Bender, Stephen J. *Venereal Diseases.* Dubuque, Iowa: Wm. C. Brown, 1971.

Eschenbach, David A., and King K. Holmes. "Acute Pelvic Inflammatory Disease: Current Concepts of Pathogenesis, Etiology and Management," *Clin. Obstet. Gynec.* 18, no. 1 (March 1975): 35–56.

Felman, Yehudi M. "A Plea for the Condom, Especially for Teen-agers," *JAMA* 241, no. 23 (June 8, 1979): 2517–18.

Fiumara, Nicholas J. "The Diagnosis and Treatment of Gonorrhea," *Medical Clinics of North America* 56, no. 5 (September 1972): 1105–13.

———. Lecture on venereal disease, Endocrine Training Course, Massachusetts General Hospital, 1971.

Grover, John W., and Dick Grace. *VD: The ABC's.* Englewood Cliffs, N.J.: Prentice-Hall, Inc., 1971.

Handsfield, H. Hunter. "Disseminated Gonococcal Infection," *Clin. Obstet. Gynec.* 18, no. 1 (March 1975): 131–42.

Henderson, Ralph H. "Venereal Disease: A National Health Problem," *Clin. Obstet. Gynec.* 18, no. 1 (March 1975): 223–32.

King, A., and C. Nichol. *Venereal Diseases.* Baltimore: Williams and Wilkins, 1975.

McCoy, Kathy, and Charles Wibbelsman. "Update on VD," *Glamour,* March 1980, pp. 228–32.

Morton, R. S., and J. R. W. Harris. *Recent Advances in Sexually Transmitted Diseases.* Edinburgh: Churchill Livingstone, 1975.

Nichols, Leslie. "Newer Methods of Diagnosis and Treatment of Sexually Transmitted Diseases," *J. Am. College Health Assoc.* 26, no. 6 (June 1978): 308–11.

Perine, Peter L.; William Schalla; Martin S. Siegel; Clyde Thornsberry; James Biddle; Kwei-Hay Wong; and Sumner E. Thompson. "Evidence for Two Distinct Types of Penicillinase-Producing Neisseria Gonorrhoeae," *The Lancet* 8046 (November 12, 1977): 993–97.

Porter, J. F., and P. Kane. "Sexually Transmitted Diseases," *Research in Reproduction* 7, no. 4 (July 1975).

Remington, Jack S., and Jerome O. Klein. *Infectious Diseases of the Fetus and New Born Infant.* Philadelphia: Saunders, 1976.

Rudolph, Andrew H., and W. Christopher Duncan. "Syphilis: Diagnosis and Treatment," *Clin. Obstet. Gynec.* 18, no. 1 (March 1975): 163–82.

Schacter, Julius; Lavelle Hanna; Edward C. Hill; Susan Massad; Charles W. Sheppard; John E. Conte, Jr.; Stephen N. Cohen; and Karl F. Meyer. "Are Chlamydial Infections the Most Prevalent Venereal Disease?" *JAMA* 231, no. 12 (March 24, 1975): 1252–55.

Schofield, C. B. S. *Sexually Transmitted Diseases.* Edinburgh: Churchill Livingstone, 1975.

Smith, Alice Lorraine. *Principles of Microbiology.* St. Louis: C. V. Mosby, 1977.

Thompson, Sumner E. Personal communication, 1980.

U.S. Department of Health, Education and Welfare. *STD Fact Sheet,* 34th ed. Atlanta: Center for Disease Control, 1979.

———. "Sexually Transmitted Disease Summary," Pub. no. 00-3380. Atlanta: Public Health Service, Center for Disease Control, 1980.

Wong, K. H.; R. J. Arko; W. O. Schalla; and F. J. Steurer. "Immunological and Serological Diversity of Neisseria Gonorrhoeae: Identification of New Immunotypes and Highly Protective Strains," *Infection and Immunity* 23, no. 3 (March 1979): 717–22.

Zarate, Lenore. *VD: Getting the Right Answers.* Palo Alto, Calif.: American Social Health Association, 1977.

10
Women and Alcoholism

by Marian T. Kley

We have seen that some external factors may affect the control systems that regulate the optimal functioning of our bodies. Taking oral contraceptive steroids may affect our metabolism in some cases (Chapter 7). The decision about whether or not we use them is under our control.

The occurrence of cancer in our bodies is usually not within our control. As we have seen in Chapter 8 some of us may be genetically predisposed to develop certain forms of cancer such as breast cancer. On the other hand, the likelihood of our developing some cancers, such as lung cancer, can be reduced by behavior that *is* within our control, including our decision about whether or not to smoke cigarettes.

Even though venereal diseases are pandemic in our country today (Chapter 9) we can control our involvement with them to some extent. If we know our contacts and if we know the symptoms of the diseases, we can exercise some control over the incidence of sexually transmitted diseases, at least within ourselves and our partners.

In this present chapter on women and alcoholism, and in the next chapter on diet and exercise, we shall be considering factors in our life-style that are largely within our own control. Intelligent regulation of these factors can permit our control systems to function at their optimal levels. Our regulation of our food and alcohol intake, combined with a well-chosen exercise program, is, as we shall see, a key element for putting us *in control* of our own lives, health, and well-being. The alcoholic has lost this control.

WHAT IS ALCOHOLISM?

The majority of women who drink alcoholic beverages do not become alcoholics. Nevertheless, we should know the signs and dangers of this disease so that we can protect ourselves from its consequences.

We have three definitions of alcoholism from authoritative sources. The U.S. Department of Health, Education and Welfare (DHEW) defines alcoholism as a metabolic disease which causes psychologic and sociologic problems. The American Medical Association says: Alcoholism is an illness characterized by preoccupation with alcohol and loss of control over its consumption. This will lead to intoxication if drinking is begun. Alcoholics Anonymous, a dynamic self-help fellowship long recognized as a significant alcohol treatment program, believes that alcoholism is a physical allergy that brings about a mental obsession with alcohol. This obsession leads to spiritual bankruptcy. Alcoholism is a progressive disease characterized by definite behavior patterns in its earlier and later stages.

Some of us can drink alcoholic beverages with no ill effects; for others, alcohol consumption can lead to physical and spiritual deterioration. Why is this?

Although to date there has been no single factor isolated that conclusively establishes the cause of alcoholism, most authorities agree that it is caused by an interplay of physiologic, psychologic, and sociologic factors.

Family studies show much higher rates of alcoholism among the relatives of alcoholics, particularly children, than in the general population. In addition to alcohol problems, children of alcoholics have a high frequency of antisocial behavior, neurotic symptoms, and psychosomatic complaints.

The characteristics that distinguish special population groups from the dominant culture and from each other can often be potent factors in the development of alcohol use and abuse. For this reason, youth, women, the elderly, Hispanics, American Indians, and black Americans, among others, should be considered separate sociocultural units. These groups also have unique needs for prevention and treatment of alcohol abuse, and may pass more quickly through phases in the disease of alcoholism.

WHY DO WOMEN DRINK?

For a long time, it has been assumed that women and men reach for the bottle for the same reasons and that the circumstances and conflicts that make them into alcoholics are identical. This is an assumption that is open to challenge.

Traditionally, men and women in our society have been called upon to fill essentially different social roles. The man has often assigned first priority to occupational and financial success. The woman who works as a full-time homemaker often finds that family and friends are especially important. Until recently, women who have been employed outside the home usually worked in occupations that were not so stressful to the point that something, anything, was needed to relieve the pressure. For most women, the most important disturbances predisposing them to seek relief in alcohol have been those in the area of marriage and family. It may be that the recent movement of women into more stressful job situations will change the focus of these concerns.

One reason for female abuse of alcohol is an identity crisis in middle age, characterized as the "empty nest syndrome." Women having this syndrome complain of a loss of essential life tasks, frequently precipitated by the death of a husband, extramarital relationships, divorce, or the marriage of children. These women may be extraordinarily dependent upon their husbands and children and unable to find identities or interests of their own. They are first and foremost wives and mothers. Changes in these roles and a loss of tasks lead to severe psychic stress and to loneliness. Conquest of anxiety and loneliness may be sought in solitary drinking at home. When women who drink at home, alone, recognize their dependence upon alcohol, this recognition frequently leads to a complete loss of self-esteem and still greater social isolation.

Other potential alcoholics among women come from the more highly educated among us—those who have had a college or university education and in whose social strata "social drinking" is a way of life. These women fulfill their occupational tasks without essential difficulties and so their weakness with regard to alcohol abuse becomes apparent only after some precipitating factor. Disharmony with a spouse may initiate an extra drink or two to relieve anger and resentment. This practice can lead to increasing dependence on alcohol, and eventually to chronic alcoholism.

In general, alcoholics have a more highly charged perception of their marriages than do both women and men who have no problems with alcohol. Female alcoholics often feel more unhappy in their marriages than do male alcoholics. They suffer from the fact that they are "hardly able to carry on an open and satisfactory conversation" with their husbands. Their excessive use of alcohol causes guilt feelings that oblige them to subordinate their wishes to those of their partners. Partly as a consequence, they often feel attacked and exposed by their husbands, whom they may describe as coarse and insensitive.

Many women experience marital difficulties but do not become alcoholics. Similarly, most women pass through the phase of their lives involving the "empty nest" and find other outlets for their time and energy. Why do some women try to solve their problems with alcohol?

Experience has shown that female alcoholics differ from nonalcoholics in several important psychological dimensions. First of all, the female alcoholic prefers to suppress, rather than to resolve her conflicts. She may convert them into bodily symptoms such as headaches or insomnia and then attempt to mask them with pills and alcohol. Many women who become problem drinkers try to bypass difficulties whenever possible. Out of fear of failure in social contact, they remain at home and withdraw into their loneliness. If differences of opinion arise, they usually accede.

What do women alcoholics expect of alcohol? Female alcoholics drink primarily to overcome anxiety, tension, and loneliness and in order to relax. They expect alcohol to help them escape from their personal problems and to feel more worthwhile and more feminine. The woman alcoholic expects alcohol to have a stress-reducing effect. However, excessive drinking behavior which departs from the norm leads them to especially intense feelings of guilt and inferiority. These feelings are stress producing, not reducing.

WHY *WOMEN* ALCOHOLICS?

In order to understand more clearly the problem of female alcoholism, one must determine whether specific psychic disturbances in female alcoholics are attributable to the fact that they are *women* or to the fact that they are *alcoholics.* To put it differently: Do female alcoholics differ from other women in characteristics besides those of their alcoholism? Do they differ from male alcoholics in features other than sex? Do they differ from male alcoholics in a way that is qualitatively different from the way in which "normal" women differ from "normal" men? Finally, does there exist one specific feature or even several features that explain all these possible differences?

At the present time our answers to these questions are limited. They come from two sources—a study of alcohol effects in women during the menstrual cycle (Jones and Jones, 1976) and from the experiences of those who counsel female alcoholics.

We have seen (Chapter 7) that studies of the effects of female hormones on behavior have been inconclusive. However, Jones and Jones studied the effect of a well-defined substance, alcohol in measured amounts, in women who were in different phases of their menstrual cycles. These women, whose mean age was 25.7 years, and whose mean cycle length was 28.8 days, characterized themselves as moderate social drinkers.

The women who were tested during the premenstrual phase of their cycles reached significantly higher peak blood alcohol levels than women tested at other times in the cycle. When body weight was taken into consideration, women also obtained significantly higher peak blood alcohol levels than did males at the equivalent dose of alcohol.

During the premenstrual phase of the menstrual cycle, estrogen and progesterone levels begin to fall rapidly (as was illustrated in Figure 4-6) and it may well be that there is a direct association between hormone levels and the degree to which our bodies are affected by alcohol. If this is the case, postmenopausal women who do not take estrogen may be more vulnerable to the effects of alcohol. The relationship between hormone levels and alcohol tolerance in women could be greatly elucidated by further research.

The other specific risk factor for female alcoholics, which cannot be explained by hormones, is a marriage in which the female alcoholic perceives her spouse as critical and insulting, in which she must submit to his demands, and in which she shuts herself off from her partner. In this situation, a woman may use alcohol to experience a sense of compensation for personal deficits. Following alcohol consumption she experiences briefly feelings of enhanced social activity and security.

These direct positive consequences reinforce the drinking and maintain it. Over the long term, drinking to an extent that deviates from the norm leads, especially in female alcoholics, to massive feelings of guilt. Since guilt feelings do not emerge until later, they fail to have a corrective effect upon drinking behavior. Instead, they provide a further basis for feelings of inferiority, helplessness, and stress, and, hence, can become an additional triggering stimulus for renewed drinking. The female alcoholic may then pass from the early alcoholic phase to the crucial phase or from the crucial phase to the chronic phase unless she receives some form of help or therapy.

STAGES IN THE DEVELOPMENT OF ALCOHOLISM

The gradual loss of control and progression of the disease of alcoholism is summarized in Table 10-1. The phases in this progression are explained below.

Prealcoholic Phase

The prelude to alcoholism is a time when social drinking is intermittent, when drinking is done at parties or to celebrate an occasion. And on those occasions, one cocktail or so suffices. During this phase we drink only for social reasons—there is, as yet, no psychological or physiological "need" to drink.

TABLE 10-1 Progression of the Disease of Alcoholism

Prealcoholic	1. Uses alcohol as a drug. 2. Drinks larger amounts progressively.
Early alcoholic	1. *Blackouts.* Behavior is controlled by the subconscious and there is no recall of such actions, although to others action seems natural. 2. *Sneaking drinks.* 3. *Preoccupation with drinking.* Absorbed in how good a drink will taste. 4. *Guilt.* Realizing drunken behavior was unacceptable. 5. *Alibi System:* a. *Denial.* "I only had two beers." b. *Rationalization.* "I wouldn't have had that drink, but he said . . . he looked . . . he did. . . ." c. *Projection.* "I won't take a drink until five o'clock." d. *Fragmentation.* The feeling that hands, feet and the entire body are about to explode and fly off in different directions.
Crucial phase	1. *Loss of control.* 2. *Grandiose behavior.* Defensive behavior, in the form of dignified, snide remarks. 3. *Marked aggressive behavior.* Sometimes vitriolic sarcastic remarks. 4. *Periods of total abstinence to prove control.* 5. *Change of drinking pattern.* From whiskey to vodka. 6. *Change of associates.* Generally, to those among whom drinking behavior is accepted. 7. *Loss of friends.* 8. *Withdrawal syndrome.* Drinking at home, alone.
Chronic phase	1. *Benders.* 2. *Drinking with the lower class.* Or alone. 3. *Ethical deterioration.* Discarding principles of honesty, fidelity; retreating into lying to self and others. 4. *Loss of alcohol tolerance.* Actually unable to stop drinking and/or taking psychotropics of any kind.
Death or insanity	

1. Uses Alcohol as a Drug The phrase "uses alcohol as a drug" implies that we use alcohol for psychological relief—to relax us, to put us in a better frame of mind, or because we feel we are "entitled" to a drink after a hard day at the office, with the kids, or doing housework. Often in the late afternoon, before the busy time of preparing dinner, we sit and relax with an alcoholic beverage of some kind (beer, wine, whiskey, or vodka).

2. Drinks Larger Amounts Progressively As time goes on, one drink leads to more or to larger drinks. As some people drink more, they lose control of their drinking, and they may drink alone. Drinking alone is a sign that points to the possibility of alcoholism.

FIGURE 10-1 The National Institute of Alcohol Abuse and Alcoholism (NIAAA) estimates that one out of every 30 women in the United States is an active alcoholic. Cited by James H. Winchester, "The Special Problems of Women Alcoholics," *Reader's Digest,* March 1978, p. 207.

Early Alcoholic

Steps 1 through 5 do not always occur in the order listed in Table 10-1. Their order is less important than the behavior patterns themselves and the realization that alcohol is controlling those patterns.

1. Blackouts Blackouts are those times when we don't remember what happened for an hour—or a day. One woman reported that she drove her car for six days through two states and had no recollection of anything that happened. She knew she had been in five towns in those six days because her credit card had entries in those towns. She drank during the entire time, but ten years later she has not been able to recall her actions during that time. The only thing she is sure of is that her subconscious drove her car and registered at motels, and that she signed for the charges with an acceptable signature. Blackouts can occur without alcohol or any other drug, but generally these blackouts last only for a few seconds. For instance, we find ourselves standing in front of the refrigerator and for a moment we wonder what we need there. In seconds we recall what it is and that's the end of the blackout. Alcoholic blackouts are far more serious, not only because they last longer, but because our faculties may be dangerously impaired while we are drinking.

2. *Sneaking Drinks* While visiting in a friend's home we offer to pour the drinks and, while alone in the kitchen, we quickly pour ourselves a drink and drink it, pouring another to put on the tray to take back to the party. Or we hide a bottle in some secret place other than the liquor cabinet so we will be assured of a drink when we want (need) it.

3. *Preoccupation with Drinking* Preoccupation with drinking can take many forms, such as: (a) Anticipating how a drink will taste and how it will make us feel (relaxed . . . exhilarated . . . convivial) while we are engaged in some office work or homemaker chores; (b) planning on buying a bottle after work; (c) thinking who will be at the bar when we get there; or (d) deciding ahead of time what kind of a drink we will make when we get home.

4. *Guilt* Guilt isn't always recognized as such, but most of us who are excessive drinkers know we shouldn't drink so much, intuitively know we are headed for trouble. The saying "Alcoholism is a suicide disease" is true. The alcoholic recognizes long before others do that alcohol is a problem and is getting progressively worse. The alcoholic feels guilt when she wakens in the morning with a hangover and vows to stop drinking *so much.* But until vows are made to stop drinking *entirely,* the slide into acute alcoholism continues, gathering speed on its downward course. Guilt prompts us to deny we have had five drinks, only owning up to two. Of course we only fool ourselves, not anybody else.

5. *Alibi System* The alibi system is a familiar behavior pattern for any drinking alcoholic. It can take a number of forms, including denial, rationalization, projection, or fragmentation.

Denial. The alcoholic who uses denial refuses to admit the extent and seriousness of her drinking. When asked by husband, sons, daughters, friends, "Have you been drinking? You said you were going to stop," she replies "I only had two." Nobody but the alcoholic is fooled.

Rationalization. Rationalization means going through the intricate system of concocting "valid reasons" for drinking: "I only needed a drink because . . . (fill in any vague or traumatic trumped up reason). However, to others such "reasons" turn out to be excuses that are better left unsaid.

Projection. Signs of projection include the ability to actually *taste* the drink that we will have after five o'clock when it's time to go home from the office. Or the drinks that we can have at home when this dull party is over. The alibi, "But I'm not drinking at the moment," serves as a poor cover-up for the fact that we are obsessed with the need for a drink and plan to fulfill that need as soon as possible.

Fragmentation. Fragmentation is that feeling the morning after the night before when we feel our head and arms and feet are going to fly off in separate directions and sunlight, or even daylight is too painful to bear. Of course, a drink will cure this helpless panic. Unfortunately, one drink is never enough.

Crucial Phase

1. Loss of Control When we have lost control, we no longer have a choice—we must drink to live. Housework, dressing to go to work, even attending to the simplest task cannot be accomplished without "just one drink."

2. Grandiose Behavior Grandiose behavior includes bluffing it out at a party, at a bar, or at home with our husband. We stand tall, look down our nose at the other person, and say (aloud or through our body language) we are going to do what we want, "and no flack, please."

3. Marked Aggressive Behavior Aggression does not necessarily mean physical violence. Sarcasm is aggressive; ridicule and insults are aggressive; tapping our foot or fingers while the other person is talking is aggressive. A demand to have the other person behave according to our dictates can be communicated by body language that is aggressive even though we never touch the other person. Such behavior is caused by a deep inner insecurity. Aggression is a defense mechanism that serves only temporarily to protect us against acknowledging this insecurity.

4. Periods of Total Abstinence to Prove Control During periods of self-imposed abstinence we try to prove to ourselves and others that our drinking problem has not taken over our life, that "I can stop any time I want." This is a deadly trap, because we are lulled into thinking our alcohol problem is under control at last. But the effort to deal with emotional problems has not been exercised and, therefore, when anger, self-pity, resentments, and envy once more take control, we know what will ease the pain. Alcohol will deaden the inner conflict—for a while.

5. *Change of Drinking Pattern* We may switch from wine to beer or from bourbon to vodka, thinking this change will somehow alleviate our drinking problem. We are unwilling to face that the problem is alcohol and that this toxic chemical is the basic ingredient of all of these beverages. We may confine our drinking to a certain time of the day or day of the week, but we never admit that drinking alcohol is the source of our problems.

6. *Change of Associates* When people we know and who know us begin to suggest that perhaps we drink too much we simply find new friends who don't care how much we drink. The saying that "water finds its own level" applies to the problem drinker. If glances of disapproval come our way from friends or loved ones, we just find new friends who don't disapprove of our behavior or we don't drink in public anymore. We drink at home, very often alone. We become "closet drinkers."

7. *Loss of Friends* Those friends we ran from may wonder what happened to us. Because we so often failed to keep appointments, insulted people, or "passed out" at parties, they may even be glad they no longer have to put up with the erratic behavior of a drunken woman.

8. *Withdrawal Syndrome* When our drinking has alienated us from family and friends, we drift away from love and understanding. We tend to withdraw from all social contacts and sink into the quagmire of self-pity and hopelessness.

Chronic Phase

1. *Benders* Trying to cope with isolation, we drink more and more until, realizing the condition we're in, we stop for a while. But life is so complex and our nerves are so ragged we decide to take a drink to get some peace. It is impossible by this time to drink one or two, so we drink to saturation and the next day we are once more in the valley of despair.

2. *Drinking with the Lower Class* The behavior of a drunken woman is more shocking to people of our culture than that of a drunken man. When we, as women, stagger around drunk in a fashionable cocktail lounge, we are taken home, asked to leave, or the police are called. The next time we drink we give some thought to going to a bar where our behavior won't generate so much criticism. A lower-class bar is one of the way stations on the road to alcoholism.

3. Ethical Deterioration Often we wake up the next morning, whether at home or in some strange bed, confronted with the fact that the principles we have been taught from childhood have long since disappeared. This realization is a very private one and may be instrumental in turning us around. Realizing we have lost our self-respect is an important factor in the decision to stop drinking. When we can no longer find any redeeming characteristics on which to pin some self-esteem, we must recognize that we have come to the last sign post. We have reached the end of the road to alcoholism. No matter how this recognition comes about, it is vitally important as the first step to sobriety.

4. Loss of Alcohol Tolerance Knowing that we drink to live and live to drink shatters us, but we no longer care. The only thing that is important to us is that we don't run out of booze. This fear rides herd on all other emotions. At long last the myth that we could stop when we wanted to—that fantasy that we believed for so long—has been exploded. We cannot stop drinking. And so we die.

THERAPY

Therapy for female alcoholism must be applied at various levels of behavior. Therapeutic intervention must aim both at a change in social behavior and also at a concrete change in drinking behavior. A change in social behavior alone does not automatically effect control of excessive drinking. Isolated treatment of the drinking problem would leave unaccounted for all those triggering stimuli that have been important contributing factors to alcohol abuse.

Alcoholics experience and expect of alcohol important positive consequences such as self-confident behavior in social situations, greater activity, and the reduction of stress and tension. However, with increasing duration and quantity of drinking, these expectations remain not only unfulfilled but are converted into failure to reach the goal. Advanced alcoholics still become more uncertain and anxious, but have no memory of these deteriorations because of "blackouts."

The detoxification of the alcoholic may appropriately utilize sedative medication, but psychotropic drugs should be prescribed with great care, if at all. The *long-term* management of alcoholism (a type of sedative drug abuse) does not ideally involve prescribing another sedative drug. Most physicians involved with large outpatient alcoholism treatment programs agree that prescription sedative drugs are an almost total obstacle to recovery for the alcoholic. Recent findings indicate that among the 60 percent of a population of alcoholic women who had abused psychoactive drugs, three quarters had obtained them from physicians during concurrent abuse of alcohol or other drugs. Prescribing drugs to such women is very dangerous in light of the increased rates of depression and suicide attempts among alcoholic women who are polydrug users.

This is not to say that the woman alcoholic will not *want* sedative drugs prescribed for her. Many will specifically request them and will be very skillful in manipulating the doctor into prescribing them.

Although drugs are best avoided, if they are used, careful consideration should be given to the use of hydroxyzine, phenobarbital, or the benzodiazepines. The patient must be seen daily and provided with only enough medication *to last the subsequent 24-hour period.* Among the dangers in prescribing sedative drugs for alcoholics are the following: (1) Since alcoholic patients are already prone to be dependent upon their favorite sedative-hypnotic drug (alcohol), they are susceptible to dependence upon other sedative-hypnotics. Just as with alcohol, they can quickly build up a tolerance to larger and larger doses of any other sedative-hypnotics. (2) Sedative pills tend to reinforce the notion in the patient's mind that if there is a problem—that is, if she is nervous or upset—the answer lies in putting something into her mouth to make her feel better. (3) Any intake of tranquilizers or any other sedative drug isolates the patient from a very important source of help: Alcoholics Anonymous. This self-help treatment depends on total abstinence from mood (and mind) altering drugs.

Experience has shown that alcoholics do very poorly on diazepam (Valium). Many alcoholics are unaware that their faculties of judgment and coordination may be impaired by the simultaneous use of these drugs. Accordingly, those who use alcohol and diazepam simultaneously should be informed of the danger resulting from combined use, particularly regarding *driving, using sharp instruments* at work, and other potentially hazardous activities.

HIDDEN PROBLEMS

Female problem drinkers who are homemakers can easily hide their alcoholism. Compared with adult male alcoholics, their intake of alcohol is less, but they are more likely to use other drugs concomitantly.

Alcoholic women are likely to come from families with increased rates of alcoholism; also, they are more likely than alcoholic men to select alcoholic mates. It is not surprising, then, that alcoholic women have an increased incidence of sexual problems, although there is a serious question as to whether the alcoholism is a cause or a consequence of these problems.

Conservative estimates of the number of adult women with alcohol problems range from 1.5 million to 2.25 million. Among women, social factors are predictive of drinking practices. For example, under age 35, those who are divorced or separated have the highest incidence of problem drinking. Among married women under 65, those who are working have higher rates of problem drinking than those who are not employed outside the home, regardless of socioeconomic status. However, we are reminded that *women who are not employed and who have drinking difficulties remain overlooked by society.*

Alcoholic women very often begin their heavy drinking later in life and ingest smaller quantities than do alcoholic men. Therefore, they show less evidence of physiologic damage due to alcohol use. This makes it difficult for a physician to identify alcohol abuse in a female patient, especially when her family and friends facilitate her drinking by denying the existence of the problem. If she is a homemaker and if family and friends enable her to hide her drinking, the alcoholic woman is unlikely to confide in her physician regarding alcohol use. In time, however, alcohol will affect nearly every system in her body.

PHYSIOLOGICAL EFFECTS OF ALCOHOL CONSUMPTION

Alcohol has a pervasive effect on the body from its absorption through the gastrointestinal tract and its transportation in the bloodstream. The brain and central nervous system, the heart, the muscles, the endocrine system, and, in pregnant women, the embryo and fetus are all affected.

Alcohol is associated with muscle deterioration, including deterioration of the heart muscle. Diseases of the coronary arteries, such as angina pectoris and myocardial infarction, increase with heavy alcohol consumption. Statistics show that the disease of alcoholism shortens life expectancy by an estimated 10 to 15 years.

The report of the World Health Organization (1977), cites evidence indicating that alcoholism has two components, each of which is a significant problem in its own right: (1) the *alcohol dependency* syndrome, a specific illness; and (2) *alcohol-related disabilities*—social, mental, and physical. A person with alcohol-related disabilities may or may not be dependent on alcohol. Excessive focus on identifying alcoholic-dependent persons may have diverted our attention from the damage that alcohol can do to the user who is not actually dependent on it physically or psychologically. The WHO report stresses that *the alcohol dependency syndrome and alcohol-related disabilities should be diagnosed independently.* The National Institute of Alcohol Abuse and Alcoholism (NIAAA) endorses this view: Treatment programs must be concerned both with alcohol dependency and with alcohol-related disabilities.

Alcohol-related disabilities and their causes among adults and youth are summarized in Table 10-2.

TABLE 10–2 Alcohol Related Disabilities and Their Causes

Disorder	*Cause*
Hypoglycemia	Inhibition of glucose formation, depletion of glycogen stores, low carbohydrate diet
Hyperlipidemia	Increased lipoprotein production, increased lipoprotein clearance, mobilization of nonhepatic fat stores
Hyperuricemia	Decreased renal clearance of uric acid
Esophagitis	Direct toxic effect of alcohol (vomiting)
Gastritis	Increased secretion and histamine production, direct toxic effect of alcohol
Duodenal ulceration	Increased secretion and histamine production
Steatorrhea and malabsorption	Pancreatic insufficiency, thiamine deficiency, hyperperistalsis
Fatty liver	Triglyceride accumulation in liver cells
Alcoholic hepatitis	Inflammation and necrosis in liver cells
Cirrhosis	Scarification of liver parenchyma
Pancreatitis	Increased secretin production, malnutrition
Various anemias	Direct toxic effect, malabsorption, malnutrition
Beriberi heart disease	Thiamine deficiency
Cardiomyopathy	Direct toxic effect of alcohol and malnutrition
Skeletal myopathies	Direct toxic effect of alcohol

LONG-TERM TREATMENT OF THE FEMALE ALCOHOLIC

It is necessary to understand both physiological and psychological reactions to alcohol in our bodies in order to start the treatment process. The *starting point is total abstinence from alcohol.* Alcohol is not harmful to most people who drink it. But it is an addictive drug to the minority of the population who are alcoholics. The exact difference in body chemistry between the alcoholic and the non-alcoholic has not been ascertained. However, long before the chemically imbalanced people—that is, the alcoholics—became problem drinkers, their reaction to alcohol was noticeably different from that of nonalcoholics.

The blood sugar level of potential alcoholics was probably low *previous* to problem drinking. Low blood sugar has various negative emotional reactions. Alcohol raises the blood sugar level to normal, sometimes even abnormal levels. This gives the drinker a sense of well-being, an ability to cope with her problems. Between bouts with alcohol her blood sugar level drops and she often shows signs of hypoglycemia.

The term "orthomolecular psychiatry" was coined by Linus Pauling to describe the effects of the body's physical condition on the mind and behavior. Orthomolecular psychiatrists believe that factors like low blood sugar, deficiency in vitamins B_6 and B_{12}, caffeine dependency, and alcoholism can all cause criminal behavior. Hypoglycemia is a metabolic disorder characterized by feelings of dizziness, exhaustion, depression, and antisocial behavior. What is the relationship between low blood sugar (hypoglycemia) and alcoholism?

A high percentage of the people who are known alcoholics or who are intermittently violent reported symptoms of hypoglycemia and corrected the problem by switching to a sugar-free diet of lean meat, fresh fruits, and vegetables. As the hypoglycemia was corrected, so was the antisocial behavior. On the other hand, when actual intake of alcohol has ceased (the morning after heavy drinking), blood sugar level drops, sometimes far below normal, and the drinker often instinctively begins to think about another drink. If she does not have the drink, her blood sugar sinks lower and she suffers acute depression, headache and/or nausea. Very often if the blood sugar level is at an extremely low point for a long period of time, the alcoholic becomes violently destructive.

Smoking cigarettes may also affect the blood sugar. Within a short time after finishing a cigarette, slight nervousness is felt. This is caused by a drop in blood sugar level. This may be one reason why one might "need" a cigarette soon after the last one has been put out. If dependence on nicotine is such that one grows faint, tired, or nervous when not smoking, chances are that low blood sugar plays an important role in this depressed feeling.

Williams (1959) has proposed a theory that links alcoholism with low blood sugar. He maintains that the "dry jitters" which affect recovering alcoholics are a direct result of low blood sugar. The practice of some recovering alcoholics, to carry a candy bar or some other sweet in order to relieve these attacks, is cited as evidence of their blood sugar problems.

Williams believes that in dealing with hypoglycemia we need to keep in mind not the point *from which* the blood sugar level falls or the point *to which* it falls, but rather, the *suddenness of the fall.* This sudden fall is the cause of the nervous symptoms that cause so much trouble for the alcoholic (the need to seek another drink, *fast!*) or the recovering alcoholic ("dry jitters"). A high-protein, moderate-fat diet tends to alleviate the rapid plunge of the blood sugar from one level to another. Both protein and fat are assimilated slowly and their end products of digestion reach the cells of the body gradually (see Chapter 11).

In spite of the fine work that is being done by Alcoholics Anonymous and other groups, some alcoholics suffer a relapse. Following a period of abstinence, the alcoholic appears physiologically normal, and yet she returns to drinking. Tintera (1966) believes that anyone who has not corrected the basic physiological difficulty (low blood sugar) cannot remain sober:

The aim of the low blood sugar treatment is restoration of homeostasis for all the endocrine factors involved. Therapy consists chiefly of

TABLE 10-3 Recommended Diet for Hypoglycemia

Foods Allowed	*Foods to Avoid*
All meats, fowl, fish, and shellfish	Potatoes, corn macaroni, spaghetti (made from white flour), rice, cereals
Dairy products (eggs, milk, butter, and cheese)	Pie, cake, pastries, sugar, candies
All vegetables and fruits not mentioned below	Dates, raisins, and other dried fruits
Salted nuts (excellent between meals)	Cola and other sugared soft drinks
Peanut butter, oats, and jerusalem artichoke bread	Coffee and strong tea
Gelatin with whipped cream.	Alcohol in all forms
Decaffeinated coffee, weak tea, and sugar-free soft drinks	
Soybeans and soybean products	
Oatmeal	
Certain macaroni and spaghetti, made with unbleached unprocessed whole wheat flour	

Source: Modified from Roger J. Williams, *Alcoholism: The Nutritional Approach* (Austin: University of Texas Press, 1959); and John W. Tintera, "Stabilizing Homeostasis in the Recovered Alcoholic through Endocrine Therapy: Evaluation of the Hypoglycemic Factor," *Journal of the American Geriatric Society* 14, no. 2 (February 1966): 126–50.

the administration of adrenal cortex extract, adequate nutrition and psychologic guidance. Diet is of extreme importance. Initially a rather high fat content is allowed, but eventually the diet should be high in protein, moderate in fat, and low in readily available carbohydrates.

The recovering alcoholic should be informed of specifics with regard to low blood sugar. She must be warned of possible reactions if she doesn't eat frequently (ideally, five small meals a day) to maintain her blood sugar. Low blood sugar may be a common denominator between obesity and alcoholism. To deter the jitters caused by low blood sugar, one person may take to soft drink or candy binges, while another will turn to alcohol. If the first kind of person does not correct the blood sugar condition she is likely to become fat. The other kind of person is destined to become an alcoholic. A diet that may be helpful to both types of women can be found in Table 10-3.

FETAL ALCOHOL SYNDROME

In recent years scientists have begun to investigate the validity of an ancient belief that women who drink heavily during pregnancy significantly increase their risk of bearing deformed and mentally retarded babies. The Greeks prohibited the use of alcoholic beverages by newly married couples so that defective children would not be conceived.

Evidence from animal studies (Brown et al., 1979) and observations of the children of women alcoholics (Streissguth et al., 1980) have been used to alert all of us, particularly pregnant women, to the long-term consequences of the Fetal Alcohol Syndrome.

The excessive use of alcoholic beverages during pregnancy produces infants with abnormalities. Developmental delay (mental retardation) is the most serious consequence of the Fetal Alcohol Syndrome. Although improving the home environment might be expected to help this condition, the positive influence of environment is minimal for these organically damaged children.

Motor dysfunction (lack of coordination of the arms and legs) is common among children who start life with the handicap of Fetal Alcohol Syndrome. As infants, they have difficulty in sucking. They have a slight tremor that causes their hands and heads to shake, and often causes the whole body to twitch. They have little or no strength in their hands and fingers to pick things up, as well as poor eye-to-hand coordination.

Sometimes these children engage in self-stimulating behavior such as rocking (head-knee position in crib or bed, rocking up and down), head banging and head rolling. Often they are hyperactive and/or have very short attention spans.

The most common facial defects are shortened muscles in the eyelids and a decreased eye growth, so that the child must tilt its head in order to see. A receding upper gum may give the effect of no upper lip.

The abnormalities listed above as the Fetal Alcohol Syndrome are not seen in the infants of all chronic alcoholic women, nor are all children who have these abnormalities necessarily born of chronic alcoholic mothers. However, when clinical records were kept on 23 chronic alcoholic women, one-third of these women gave birth to infants with complete Fetal Alcohol Syndrome defects. One-half of the children had some of the defects and 17 percent of the infants died prior to birth.

Studies of the Fetal Alcohol Syndrome have resulted in the recommendation that from the moment a woman is aware of her pregnancy she should consider the seriousness of over indulgence in alcoholic beverages. Approximately 80 percent of the alcohol consumed is detoxified by the liver, but the lethal 20 percent balance travels through the body via the bloodstream and into the placenta. This alcohol surrounds the fetus; washing over the brain and central nervous system every 30 seconds.

The suggested rule of thumb for pregnant women is to limit daily alcohol drinking to one (and *only* one) of the following:

two mixed drinks (one ounce distilled alcoholic beverage per drink), or

two 5-ounce glasses of wine, or

two 12-ounce cans of beer.

The risk factor to the fetus increases with each drink after the limited two drinks. The two drinks should be spaced at least an hour apart because it takes the liver this long to detoxify the alcohol. It is recognized that as many as six drinks per day may seriously harm the fetus.

Women of childbearing age need to remember that everything we eat, drink, and do affects the fetus. Prenatal care lasts a lifetime—the lifetime of our children.

THE PROBLEM WITH ALCOHOL: ADMITTING IT AND SOLVING IT

A pregnant woman who uses alcohol has added motivation to control her consumption for the duration of her pregnancy. What does a woman do when she, or someone close to her, realizes that she has a problem with alcohol? The motivation to control alcohol consumption or to relinquish it altogether must be found. Where and how?

Many women are reluctant to consult a physician when they are sober, or even partially sober, unless they are seeking relief from the pain of sobering up. Of course when a crisis situation occurs—such as convulsions, D.T.'s, and/or falling into a comatose state (passing out) from which family or friends are unable to arouse them, the physician is called. But the woman alcoholic rarely calls or visits a physician except under pressure from family or employer. After she is rescued from the crisis condition (both physical and psychological) her ongoing alcohol problem is seldom discussed.

Confrontation with the results of excessive drinking may be helpful to the alcoholic who cannot admit that she has a drinking problem. When in the state of euphoria while drinking, the alcoholic considers all of her problems surmountable ... some time in the future. And with a hangover, her insurmountable problems are too great to try and solve. At that time, of course, her primary problem is where the next drink is coming from.

At some time between the euphoria of intoxication and the depression of a hangover the woman alcoholic needs to see her problem and to recognize herself someplace in the chart on Table 10-1. She herself must decide to regain control over her own life, which alcohol has made unmanageable.

When she has made the decision to stop drinking alcohol, she will need help and support. In order to stop drinking, initially, she may need a medication that will make drinking alcohol undesirable and even dangerous.

Disulfiram (Antabuse) should be prescribed only with the patient's (and the family's) full knowledge and consent. The contradictions to disulfiram administration are these:

1. *If a person has brain damage or mental deficiency to such a degree that she cannot comprehend the significance of the instructions, she should not take the drug.*

2. *If a person is strongly suicidal, prescribing disulfiram would provide another potential means of committing suicide.*
3. *If a person manifests an allergic reaction to the drug, it should not be prescribed.*
4. It should not be given to anyone with alcohol in her blood.
5. *Severe coronary artery disease should be at least a strong relative contraindication, as most of the deaths in those who have drunk alcohol while taking disulfiram have occurred in those with prediagnosed coronary artery disease.*
6. *There are four medications that a person should not take while on Antabuse: isoniazid (INH), sodium warfarin, hydantoin, paraldehyde, and metronidazole.*

Antabuse stays in the body at least several days after the last dose. Alcohol-Antabuse reactions have been experienced as long as two weeks after the last pill has been taken.

The alcoholic who is taking Antabuse to help her stop drinking knows that drinking any form of alcohol during the time she is taking this medication will make her violently ill. Some patients need this motivation to abstain from alcohol so that they can begin recovering from the disease. The recovering alcoholic may also need special help during the alcohol withdrawal period.

The severity of the alcohol withdrawal syndrome depends upon at least six factors:

1. *The amount of alcohol consumed.*
2. *The duration of the drinking episode.*
3. *The individual's previous experience with alcohol withdrawal. Those who have previously suffered convulsions or delirium tremens are much more at risk of experiencing them again.*
4. *Coexisting disease problems. Other things being equal, a patient with pneumonia is more likely to suffer a severe alcohol withdrawal syndrome.*
5. *Nutritional factors. Coexisting deficiency states tend to aggravate the condition.*
6. *Age. Elderly persons will have a more severe withdrawal syndrome.*

Abstaining from alcohol is only the first step in recovery from acute alcoholism. Nutritional factors play an important role in the withdrawal and recovering process. A high-protein diet, vitamins B_1, B_3, B_6, and B_{12}, folic acid, calcium, and magnesium are highly recommended to restore good health. It must be remembered that vitamins do not reach full potential if they are not used in conjunction with minerals and the amino acids contained in protein. In European clinics, vitamin B_{12} is used in massive doses during early recovery from alcoholism. The combination of magnesium, calcium, and niacin will often relieve the painful leg cramps experienced by many alcoholics going through the early stage of recovery.

It is important to remember that all three definitions of alcoholism we have given include physical, psychological, and sociological problems, in that order. All three make up the composite illness called alcoholism. We are not built in separate compartments—physiological, sociological, and psychological. These are all parts of our lives, and alcoholism has its roots in all three. For this reason, involvement with Alcoholics Anonymous (AA) may be the best treatment available for any alcoholic. Psychiatrists, psychologists, and physicians offer help to the person afflicted with the disease of alcoholism. But Alcoholics Anonymous offers a *way of life* without alcohol. It is responsible for the recovery of more alcoholics than any other treatment modality. Doctors, counselors, and mental health workers should make every effort to utilize this key resource.

Only other alcoholics and members of Alcoholics Anonymous truly understand the emotional stress of the alcoholic. This worldwide fellowship has groups in most cities and is listed in telephone directories. It is based on the principles of honesty with oneself, open-mindedness, willingness, and a belief in the unfailing assistance of a Higher Power. Members share their experience, strength, and hope with each other through group therapy. The disease of alcoholism is their common bond. In some cities, special AA groups made up entirely of women meet regularly to help each other maintain control over their lives—without alcohol.

The belief in education as a solution to social problems is longstanding in the United States. Education in the field of alcoholism is no exception. Despite the lack of a body of evaluative literature to indicate its effectiveness, alcohol education prevails as a major tool in the attempt to prevent alcohol abuse. As we learn more about the disease of alcoholism and its prevalence in our society we can decide for ourselves whether we will control the use of alcohol in our lives or whether we will allow the abuse of alcohol to interfere with this control. We can only control the effects of alcohol *before* we drink it, not after. If we believe strongly in mind over matter in this regard, we might try taking two tablespoons of castor oil, and then *think* "I'm not going to let it affect me." Need we say more?

REFERENCES

Abrahamson, E. M., and A. W. Pezet. *Body, Mind and Sugar.* New York: Pyramid Books, 1951.

Brown, Nigel A.; Eugenia H. Goulding; and Sergio Fabro. "Ethanol Embryotoxicity: Direct Effects in Mammalian Embryos in Vitro," *Science* 206, (November 2, 1979): 573–75.

Cheraskin, E., and Ringsdorf, W. M., Jr. *New Hope for Incurable Diseases.* New York: Arco Publishing Co., 1973.

Cohen, Sidney. "How to Become an Alcoholic," *Drug Abuse and Alcoholism Newsletter* (Vista Hill Foundation) 8, no. 8 (October 1978), pp. 2–5.

Coudert, Jo. *The Alcoholic in Your Life.* New York: Warner Books, 1974.

Cruse, Joseph; Thomas Fleming; Vernelle Fox; James R. Milam; John E. Milner; Thomas H. Clark; and Joseph Berger. "Changing Attitudes (Yours?) on Alcoholism," *Patient Care* 13, no. 4 (February 28, 1979): 15–107.

Dahlgren, L. "Female Alcoholics," *Acta Psychiatr. Scand.* 56, no. 1 (July 1977): 39–49; vol. 56, no. 2 (August 1977): 81–91; vol. 57, no. 4 (April 1978): 325–35; vol. 59, no. 1 (January 1979): 59–69.

Ewing, J. A., and B. A. Rouse. *Drinking: Alcohol in the American Society.* Chicago: Nelson Hall, 1978.

Greenberg, Leon A. "Alcohol in the Body," *Scientific American* 189 no. 6: 86–90, 1953.

Jellinek, E. M. *The Disease Concept of Alcoholism.* New Haven: College and University Press, 1960.

Jones, Ben Morgan, and Marilyn K. Jones. "Alcohol Effects in Women during the Menstrual Cycle," *Annals of the N.Y. Academy of Sciences* 273 (1976): 576–87.

Morrissey, Elizabeth R. "Alcohol Related Problems in Adolescents and Women," *Postgraduated Medicine* 64, no. 6 (December 1978): 111–13, 116–19.

Streissguth, Ann Pytkowicz; Sharon Landesman-Dwyer; Joan C. Martin; and David W. Smith. "Teratogenic Effects of Alcohol in Humans and Laboratory Animals," *Science* 209 (July 18, 1980): 353–61.

Tintera, John W. "Stabilizing Homeostasis in the Recovered Alcoholic through Endocrine Therapy: Evaluation of the Hypoglycemia Factor," *Journal of the American Geriatric Society* 14, no. 2 (February 1966): 126–50.

Williams, Roger J. *Alcoholism, The Nutritional Approach.* Austin: University of Texas Press, 1959.

Winchester, James H. "The Special Problems of Women Alcoholics," *Readers' Digest.* March 1978, pp. 207–12.

World Health Organization, "Alcohol Related Disabilities," Technical Report Series no. 32. Geneva: World Health Organization, 1977.

11
Diet and Exercise
by Carol Ann Dyer

In earlier chapters we have seen how our lives are shaped by our genetic endowment and how our genes regulate the control systems of our bodies. We have become aware, also, of choices we may make that affect our inborn control systems, such as the use of oral contraceptive steroids to regulate fertility or estrogen replacement medication to offset some undesirable physical effects of the menopause. We know about the consequences that our choices to use alcohol and tobacco may have on our health. Although it may not be obvious, choices we make about diet and exercise have a profound effect on our lives as well.

FOOD: A MATTER OF CHOICE

The authors of a popular American cookbook begin their book this way:

We enjoy the cynical story of the old-fashioned doctor who insisted first on going straight to the kitchen of the afflicted household. Not until he had effusively thanked the cook for giving him a new patient did he dash upstairs to see how he could relieve the cook's victim. The fact is that everyone who runs a kitchen can, in the choice and preparation of food, decisively influence family health and happiness. [*Rombauer and Becker, 1964*]

The choices we make about the foods we eat are highly personal choices, and they are influenced by many factors. Food fulfills basic social needs. Weddings may be celebrated with feasts, birthdays with special dinners. Friends are invited to share meals. Group outings to fast-food restaurants are a firmly established adolescent ritual. Business deals are negotiated at lunch. There is a significant emotional investment in food that begins in infancy as a primary contact with caregivers. There are subtle and not so subtle cultural pressures to make oneself feel better with food. Advertisers reassure us that we've worked hard and deserve to reward ourselves with their food products. The especially delicious food is that which is just like Mom used to make. Food, in short, is laden with emotional meanings that subtly (or not so subtly) affect our dietary choices.

Choosing Too Little or Too Much

Not surprisingly, because of its emotional associations, food can be used by some people to express their feeling states. Two familiar syndromes are associated with food refusal and food overuse—anorexia nervosa and obesity.

ANOREXIA NERVOSA Most anorexia nervosa sufferers are young women. They are often the daughters of highly ambitious parents, girls who feel pressure to meet certain expectations. They also feel anxious about the feminine social role. The anorexic pattern of abstaining from food, self-induced vomiting, and frequent use of laxatives often emerges during puberty. It is said that anorexic girls are alarmed at their changing physical appearance, especially the more pronouncedly feminine thigh, hip, and breast contours. As their disease progresses, the health of these young women deteriorates, and, obviously, death can result unless there is medical and psychiatric intervention.

OBESITY There are many theories, many clichés, and many proposed "easy" cures for obesity. In most cases, obesity is caused simply by ingesting more calories than we burn up during the day. In a small percentage of cases, however, obesity may result from metabolic abnormalities. Unfortunately, many women overestimate the number of inborn causes of obesity, and those of us who should and could lose weight do not try. We hopelessly resign ourselves to being fat by saying that it's just a problem we have with our glands. Some people derisively charge that obese people have a character defect. Others speculate that shyness or lack of self-esteem cause overweight persons to hide underneath their fat. "Inside every fat person is a thin person crying to get out," is the cliché. There is some controversy about what causes obesity. Studies have shown that obese persons cannot tell if they are full or not. They respond to many food cues: the sight and smell of food, the social situation in which it is served, and the feelings and associations it evokes. Their response is not modulated by their bodies' caloric needs nor by a sense of hunger or satiety. Some theorize that babies who are overfed develop an excessive number of fat cells during infancy and these extra fat cells promote easy fat storage during later life, predisposing the individual to obesity.

There is no consensus of opinion on the best treatment for obesity. Most people do concur, however, that permanent weight control requires some equally permanent change in behavior. Any dieter knows how easy it is to regain lost pounds. Several self-help books propose easy methods of weight loss. There are support groups for dieters which usually offer more in the way of peer group contact than they do in permanent weight control. Dr. Jean Mayer, a respected nutritionist, has said repeatedly that the best way to lose weight is to reduce the intake of food, especially concentrated, high-calorie food, while continuing to eat a variety of foods. The value of this approach, as well as the health consequences of obesity, should become clear in the pages that follow.

Food Choices and Energy Requirements

Although motives and options in food choice may differ widely from person to person, nutritional requirements are not complicated. In choosing a healthy diet, we need to be aware of how much food we need and what kinds of foods to choose. Every person needs a given number of calories each day to supply energy for breathing, circulation of blood, excretion of waste products, repair of worn-out cells, synthesis of hormones and enzymes, and maintainance of a constant body temperature. The calories needed for these functions depend on the basal metabolic rate (BMR), which is defined as the number of calories burned per hour by a person at rest and expressed in terms of calories per hour per unit of body surface. The BMR is proportional to the surface area of the body. It is highest in childhood, declines steadily throughout childhood until the end of puberty, and then declines more gradually from sexual maturity until old age. A theoretical range might be 50 calories per hour per square meter of body surface for a 10-year-old girl to 32 calories per hour per square meter of body surface for a 90-year-old woman. The basal metabolic rate is lower during starvation or fasting. Certain endocrine disorders cause a lower BMR. The BMR is higher during a fever, in certain endocrine disorders, in trained athletes, and during the last trimester of pregnancy.

The activities that we choose to undertake in daily life determine how many calories above the basal metabolic requirement must be supplied in our diet. Energy requirements are well known, and charts based on these requirements may be used as a guideline (see Table 11-1).

TABLE 11–1 Energy Expenditure by a 150-Pound Person in Various Activities

Activity	*Gross Energy Cost in Calories per Hour*
A. Rest and Light Activity	50–200
Lying down or sleeping	80
Sitting	100
Driving an automobile	120
Standing	140
Domestic Work	180
B. Moderate Activity	200–350
Bicycling (5½ mph)	210
Walking (2½ mph)	210
Gardening	220
Canoeing (2½ mph)	230
Golf	250
Lawn mowing (power mower)	250
Bowling	270
Lawn mowing (hand mower)	270
Fencing	300
Rowboating (2½ mph)	300
Swimming (¼ mph)	300
Walking (3½–4 mph)	300
Badminton	350
Horseback riding (trotting)	350
Square dancing	350
Volleyball	350
Roller skating	350
C. Vigorous Activity	over 350
Table tennis	360
Ice skating (10 mph)	400
Tennis	420
Water skiing	480
Hill climbing (100 ft. per hr.)	490
Skiing (10 mph)	600
Squash	600
Cycling (13 mph)	600
Running (10 mph)	900

Source: Food and Nutrition Board, National Research Council, *Recommended Dietary Allowances,* 9th ed. Washington, D.C.: National Academy of Sciences, 1980.

It is possible to formulate a general rule of thumb about energy requirements. The cost of gasoline has made almost everyone painfully aware that it takes a lot of fuel to drive heavy automobiles at high speeds. In the same way, more calories are required for a relatively large person to exercise or work at high physical intensity than are necessary for a smaller person to work at the same intensity or for a larger person to work at lower intensity. Although it may seem inelegant to compare a human being to a machine, our bodies respond to calories with machine-like precision. Fuel, which is burned in our bodies by a process called oxidation, yields almost the same energy when it is burned in a calorimeter, an apparatus that measures heat from a chemical reaction in a laboratory. If our daily calorie intake equals the base-line requirements that correspond to our specific basal metabolism plus the requirements for whatever activity we undertake, we have a balanced energy equation and our weight remains constant. If our energy requirements exceed the yield from the food we eat that day, the body can draw upon food reserves stored as glycogen in muscles and liver, a relatively short-term storage form of food energy, or it can utilize fat stored in adipose tissue, a longer-term storage form of food energy. When these energy reserves are drawn upon over time, we lose weight. And, contrary to our most powerful wishes, if our daily food intake exceeds the energy requirements of that day, the extra food energy is stored in the body, first in the form of glycogen and then in the form of fat.

Energy from Food

Although a calorie is a calorie as far as the body is concerned, there are differences among the three basic classes of food—carbohydrates, fats or lipids, and proteins—both in terms of potential energy yield and in terms of what raw materials are made available to the body after digestion. You will remember from Chapter 7 that digestion is a process that breaks down the large complex molecules that we ingest as food. After digestion, the smaller molecular components are absorbed through the small intestine and transported to all cells via the circulatory system. These smaller molecules enter the cell and are broken down further in a process called cellular respiration. In cellular respiration, through a series of many steps, the chemical bonds that hold together the molecular products of digestion are broken. The energy that was used in making the bonds originally is released to the cell to drive the various chemical reactions that are required to maintain a human being. (Alternatively, the energy can be stored in the form of high-energy bonds in molecules within the cell and released when the energy is needed.) The smaller molecules that are the by-products of cellular respiration can be reassembled by the cell to make new molecules. If they cannot be used they are excreted as waste products. The potential energy yield of food depends on the structure of the molecules that compose it. Carbohydrates and proteins yield 4 calories per gram, alcohol yields 7 calories per gram, and fat yields 9 calories per gram. If all we needed from food were energy it would be theoretically possible, although probably monotonous, to fulfill all nutritional needs from a single source. We can begin to understand why it is necessary to have foods from all three of the basic classes if we look at the fates of the various food molecules after digestion. We shall see that carbohydrates, fats, and proteins have unique functions in our bodies, and that, while each provides calories, their functions are decidedly different.

CARBOHYDRATES: THE ENERGY YIELDERS Carbohydrates are a diverse class of food. They are named with the prefixes *mono–*, *di–*, and *poly–*, and the root word *saccharide*, which means sugar. They are classified on the basis of how many sugar units the basic molecule contains. We ingest polysaccharides in the form of starch, found in varying concentrations in fruits and vegetables, and in the form of cellulose, an essential structural component of all plants. Starch is digestible by human enzymes that help to break the bonds between the sugars that make up the molecule. The basic sugar unit is glucose. Cellulose, also a long chain of glucose molecules, is not digestible by human enzymes because the sugar units are connected in a different way. Cellulose is an important part of healthy diets, however, because the fiber contributes to a feeling of fullness and stimulates the intestine. We ingest disaccharides in the form of sucrose (table sugar), maltose (malt sugar), and lactose (milk sugar). Each of these sugars is digested by a specific enzyme and broken down into its constituent monosaccharides. Sucrose yields a unit of fructose and a unit of glucose; maltose yields two units of glucose; lactose yields a unit of glucose and a unit of galactose. Glucose, the most prevalent end product of carbohydrate digestion, and fructose are used by the cell during cellular respiration. Galactose is changed into glucose by an enzyme system in the body. We also ingest fructose directly when we eat fruits and vegetables. The principal dietary sources of carbohydrates are fruits and vegetables, bread and cereal products, and dried beans and peas. The American diet also contains a significant share of carbohydrate in the form of soft drinks, candy, baked goods, and food additives designed to enhance the stability and palatability of processed food.

Glucose is the most prevalent form of carbohydrate circulating in the blood. The level of glucose in the bloodstream is sensitively monitored and regulated by enzyme and organ systems with the result that in normal individuals there is no significant fluctuation in the amount of glucose available to the cells. After digestion, glucose (along with the other monosaccharides) passes through the membranes of the small intestine and is absorbed by the circulatory system. After it has been absorbed by the circulatory system, the level of glucose is slightly higher than normal. The blood passes through the liver where the glucose can be converted to glycogen for storage in the liver or muscles. If the liver and muscles are saturated with glycogen and the blood glucose level is still higher than normal, then a series of reactions begins in the liver to convert the extra glucose to fat.

The kinds and amounts of carbohydrates we choose to eat can have an effect on our body chemistry. Simple carbohydrates (monosaccharides and disaccharides) are digested easily, and the glucose enters the bloodstream quickly. If we eat or drink a large amount of simple carbohydrate at one time and the bloodstream is flooded with glucose, the pancreas secretes insulin to take the glucose out of the bloodstream. Unless there is another source of glucose available, our blood sugar level drops suddenly and we are left with a tired feeling. Complex carbohydrates (polysaccharides) are digested and absorbed more gradually, assuring a steadier level of glucose in the bloodstream. Including complex carbohydrates in the diet has been shown to improve the glucose tolerance of diabetics.

In addition to supplying energy needs, carbohydrates combine with other molecules to form complex molecules. These molecules have a range of functions. Glucose reacts with nitrogen-containing compounds to form amino acids, a basic constituent of protein. Carbohydrates combine with proteins to form molecules called glycoproteins, which form the ground substances of cells. Glycoproteins are components of enzymes, hormones, blood proteins, and antibodies. When carbohydrate molecules combine with lipids (fats) the resultant molecules are called glycolipids. Glycolipids are found in nervous tissue. A five-carbon sugar, ribose, and its variant, deoxyribose, are components of RNA and DNA.

If we do not ingest enough carbohydrates to supply the body's energy needs, there are mechanisms to ensure that glucose levels in the bloodstream remain adequate. The first step is to marshal the reserves of glycogen. The muscles and the liver normally store enough glycogen to satisfy a 24-hour need without glucose ingestion. The conversion of glycogen to glucose is effected by the hormone glucagon, secreted by the alpha islet cells of the pancreas, and by other hormones from the adrenal gland. After the glycogen stores are exhausted, the body's fat supply can be converted to energy sources. As a last resort, the adrenal hormones facilitate the conversion of body proteins to supply glucose, which can be oxidized to yield energy. It is virtually impossible under normal conditions to have insufficient carbohydrate in the diet, and the backup systems function to keep the supply of glucose sufficient.

In the well-known pathological situation of diabetes, there is sufficient carbohydrate in the bloodstream, but the utilization of it is impaired. As we have seen in Chapter 7, this can happen in two ways. In insulin-dependent diabetes, the beta islet cells of the pancreas do not secrete enough insulin, a hormone that is essential to promote the transport of glucose from the bloodstream across the cell membrane and into the cell. In non-insulin-dependent diabetes there is sufficient insulin, but the receptor sites for insulin on the cell membrane are not responsive. One cause of the impairment of the receptor sites is obesity. Non-insulin-dependent diabetes can be managed by controlling diet and by exercise. The dosage of insulin in insulin-dependent diabetes can be lowered by altering the diet. Diabetes is a clear case of a disease that can be caused and modified by diet. If diabetes is not treated, the body goes through all the steps to get glucose into the cell that would be followed if there were no carbohydrate in the diet; that is, it uses fats and proteins as energy sources. One consequence of this is the incomplete oxidation of fats, leading to a condition called ketosis in which ketone bodies, the by-product of incomplete fat oxidation, enter the bloodstream. Ketosis, in turn, causes a change in the pH of the blood, loss of sodium in the blood, and dehydration. These effects can be duplicated by a popular weight-loss diet that advocates no carbohydrates and high fat intake for the first week. Although the body can utilize stored fat and protein to supply its glucose needs, the most efficient utilization of nutrients depends on having carbohydrate in the diet.

FATS: THE ENERGY STORERS Fats or lipids comprise the second class of food molecules. They are made up of two kinds of molecules: a backbone of glycerol with three fatty acid molecules attached to it. Fats are named on the basis of the number of carbon atoms in the fatty acid chain, and they are classified as saturated or nonsaturated. This classification depends on the kinds of chemical bonds that link the carbon atoms in the fatty acid chain. Carbon atoms in saturated fats are linked by single bonds, while some of the carbon atoms in unsaturated fats are linked by double bonds.

We ingest fats when we eat meat, fish, eggs, dairy products, nuts, some kinds of vegetables, and vegetable oils. Fats are digested by the action of bile salts, which act as emulsifying agents, and by specific fat-digesting enzymes called lipases. The products of fat digestion are free fatty acids and monoglycerides. They diffuse into intestinal cells via micelles, the receptor sites for fatty acids, where they are reassembled into a complex particle called a chylomicron. Chylomicrons enter the circulatory system through the lymphatic system. They are carried by the lymphatic system to the venous system where they are released slowly into the bloodstream. The chylomicrons travel through the bloodstream to individual cells where they are used as energy sources. Every cell in the body except brain cells can use fats as a potential energy source. Fats yield more energy per unit weight than any other food molecule and, hence, are a concentrated source of calories.

Lipids (fats) are the most important long-term storage form of energy in the body. In addition to providing a high energy yield per unit of weight they do not attract water, as do proteins and carbohydrates, and so can be stored dry. This diminishes the amount of water that the body needs to store them and, hence, the weight it must carry. Fat stored in adipose tissue also functions to cushion internal organs from shock and as thermal insulation. Lipids in the form of wax protect the ear drum, and our skin is protected by a thin film of lipid.

Cholesterol is an infamous, ubiquitous lipid. It is a part of nerve cell membranes, and it is a precursor of bile acids, vitamin D, adrenal steroid hormones, and sex hormones. High levels of serum cholesterol have been implicated as a risk factor in coronary artery disease and atherosclerosis. There seems to be a correlation between the kinds of food we eat and the amount of cholesterol in our blood. A diet that is high in saturated fat tends to favor high levels of serum cholesterol. Certain foods themselves contain a lot of cholesterol, but it is still a matter of controversy whether dietary cholesterol influences serum cholesterol. It is clear that we do not have to eat cholesterol to have it in our bloodstream or our cells. The body can make all the cholesterol it needs from other molecules in its inventory. The fat particles that are made by the liver when there is an excess of glucose in the bloodstream have cholesterol as a component.

The body can synthesize almost all of the lipids that it needs without having them in the diet because it has enzymes that can assemble the fragments from the digestion of other food molecules. There are, however, two fatty acids—linoleic acid and arachadonic acid—that humans cannot make, and that must be taken in the diet (although arachadonic acid can be made if linoleic is in the diet). They are called essential fatty acids because their intake is essential for health. They have a role in cholesterol metabolism, and are precursors for other molecules needed by the body to maintain health. One class of such molecules is the prostaglandins, which function in a way similar to hormones as chemical messengers and regulators in the body. Symptoms of essential fatty acid deficiency are hair loss, sterility, and, ultimately, death. Fortunately, the course of the processes that lead to these conditions can be reversed if the essential fatty acids are restored to the diet. Linoleic acid is abundant in all vegetable oils except olive oil and occurs to some extent in butterfat and poultry fat.

PROTEINS: THE CATALYZERS AND STRUCTURAL COMPONENTS A third kind of food required by the body is protein. The basic units of proteins are amino acids, molecules that are made up of a nitrogen-containing group and a carboxylic acid group. Proteins are long chains of amino acids with the carboxylic acid part of one amino acid linked to the nitrogen-containing group of its neighbor. The nitrogen-containing group is called an amine group; hence, the name amino acid. The bond that holds the amino acids together is called peptide bond; thus, another name for proteins is polypeptides. The kinds of amino acids in the chain, the order in which they are linked together, and the shape the molecule has after the amino acids are linked—all determine the unique character of each protein. The structural specificity of proteins has a crucial role: Some of the hormones, enzymes, antigens, and antibodies, which are proteins, depend on their unique structure as they control important functions in the body.

There are thousands of different proteins in the human body. This great variety is achieved by using the same 22 amino acids as starting materials, but linking them in varying proportions and in a sequence that is unique to each protein, and that, in turn, imparts a characteristic shape. Proteins are part of the structure of every cell in the body. They circulate in the blood plasma as part of the immune system. They are synthesized and secreted by the endocrine system as hormones, chemical messengers that regulate and coordinate many physiological processes. Enzymes, molecules that help to digest food and to lower the energy requirements for many chemical reactions in the body, are proteins. Without proteins, life as we know it would cease.

Because every living thing, both plant and animal, has proteins, any unrefined food we eat has some protein. Nutritionists analyze the quality of protein food by its digestibility—how much of its protein is available to the person who eats it—and by its completeness—how many of the amino acids required for human protein synthesis are in the food. Rich sources of protein in the diet are dairy products, meat, fish, eggs, nuts, and legumes. Protein foods such as milk and eggs that are intended by nature to sustain young animals are the most digestible and the most complete. Meat and fish, as well, have most of the amino acids required by human beings. Vegetables, however, do not contain all of the amino acids that are needed. Therefore, vegetarians must eat a wide variety of vegetable proteins to avoid suffering from amino acid deficiencies. While one vegetable might lack a certain amino acid, another will supply it. If both are eaten at the same time, a full range of amino acids will be available.

Protein digestion is facilitated by a class of enzymes called peptidases. These break the bonds that link the amino acid units, freeing them for absorption. Young infants do not have all protein-digesting enzymes, and an infant's stomach permits some undigested proteins to pass through it. The body recognizes this undigested protein as a foreign invader, a non-self protein, and initiates an immune response to it. This response sensitizes the infant to the particular food and can be the cause of food allergies. One disadvantage of premature introduction of solid food to babies is that, once the antigen to the foreign food protein has been made, the body "remembers," and an allergic reaction may ensue each time the food is eaten. Adult stomachs do not permit proteins to pass through the membranes. Proteins are absorbed as amino acids by passing through the membranes of the small intestine, probably facilitated by a carrier molecule. They enter the bloodstream and travel to all cells in the body. Their passage into the cell from the blood is probably mediated by a carrier molecule as well.

After they enter the cell the amino acids can be used to synthesize new proteins. The DNA in the cell nucleus serves as a template, and the RNA molecules in the cell act to translate the message and assemble the amino acid fragments on the ribosomes in the cell. The kinds of proteins synthesized depend on the cell. If an amino acid is not incorporated into a newly synthesized protein, it is broken down further by a process called deamination. The amine group is split off from the carboxylic acid group, undergoes further chemical reactions, and is excreted in the urine. The carboxylic acid part of the molecule can be metabolized to yield energy.

There are no provisions for storage of amino acids in the body. Therefore, in order to supply our daily needs for protein synthesis, we must include the necessary amino acids in our diet on a regular basis. This does not mean that all amino acids must be eaten every day, but there must be a steady supply from the diet. If only a small quantity of a given amino acid is present, there is a limiting effect on protein synthesis. The cell can make protein only as long as the critical amino acid is present. The unused amino acids are broken down (metabolized) and the unusable portions excreted. Some amino acids, on the other hand, can have a sparing effect. This means that dietary requirements for certain amino acids can be reduced if others are in the diet. Some amino acids have parts and chemical bonds that the body's enzyme systems cannot make. These amino acids must be included in the diet, and are therefore called essential amino acids.

If there is not enough protein in the diet for a sustained time, some amino acids can be made available by the breakdown of tissue. Without essential amino acids, however, protein synthesis is limited and the consequences can be serious. Symptoms of protein deficiency are slow wound healing and poor conditions of nails and hair. Severe protein deficiency leads to a disease, kwashiorkor, which is prevalent among children in countries with limited food choices or resources.

Americans usually do not have to contend with limited food choices or resources. The more common phenomenon in the United States is excess protein consumption. This is the result of high levels of meat consumption, in part, and of aggressive marketing of protein supplements, a popular health food item. Since excess amino acids cannot be stored in the body, the supplements and the money spent for them go down the drain. There is no conclusive evidence that consuming high levels of protein is beneficial to health.

As females, our need for protein increases gradually and regularly from birth to puberty. Our need is highest (in the nonpregnant state) during the adolescent growth spurt, tapers off toward the end of the teen years, and then levels off at maturity. During pregnancy and lactation our need for protein increases significantly. Pregnant women store protein in their skeletal muscles from the beginning of pregnancy, and it is believed that this protein is mobilized during the later stages of pregnancy when the need for protein is highest. There is extensive development of tissue during pregnancy: Blood volume increases, the size of the uterus increases, the placenta grows steadily, and the fetus is developing as well. Adequate levels of protein consumption, along with sufficient calories to provide energy, are good insurance for a positive outcome in pregnancy. If teen-agers become pregnant, their protein intake must provide for their own development as well as that of the fetus, and they must have even more protein in the diet than adults. The amount of protein needed during lactation depends on the amount of milk that the nursing infant consumes. Obviously, if a woman is nursing more than one infant her protein intake must be increased accordingly.

Minerals and Vitamins

In addition to providing carbohydrates, fats, and proteins in the right proportions, our diets must include sufficient amounts of microscopic substances known as minerals and vitamins.

Minerals: Coenzymes, Constituents, and Components Minerals form parts of some body structures, help to control muscles and nerves, and function, as do some vitamins, as cofactors or coenzymes in the following way. One part of an enzyme, made up of protein and called an apoenzyme, is inactive. The apoenzyme is activated when it combines with a coenzyme, which can be a mineral in ionic form or a vitamin or its derivative. When the apoenzyme and the coenzyme combine, the whole entity, called the holoenzyme, is biologically active. If the needed vitamins and minerals are not in the diet and the body has no stored reserves of them, then the holoenzyme cannot be made, and the reactions in which a given holoenzyme participates cannot go to completion. All metabolic reactions are facilitated by enzymes, so the deactivation of enzymes can have significant consequences.

Calcium. Calcium is an important constituent of teeth and bones; 98 percent of the body's calcium stores are in the bones. In addition to its structural role, calcium in ionic form in the blood and in the fluid that bathes the cells helps to regulate muscle contraction, stability of membranes, and the rhythm of heart contractions. Blood could not clot if there were no calcium ions in the plasma. Neuromuscular excitability is lowered by calcium ions. Indeed, one symptom of calcium deficiency is a jumpy, twitchy feeling. A consequence of prolonged calcium deficiency can be muscles that twitch uncontrollably until they are so fatigued that they do not contract any more.

Calcium levels in the body are monitored by parathyroid hormone, vitamin D, and calcitonin. If the level of calcium in the bloodstream drops below a critical level, parathyroid hormone initiates release of calcium from the bones. Chronic deficiency of calcium in the diet can lead to a weakening of the bones. While loss of calcium from the bones can lead to osteoporosis it must be realized that increased dietary calcium alone is not necessarily a preventive factor. Vitamin D promotes the absorption of dietary calcium from the intestine. Calcitonin opposes the action of parathyroid hormone; that is, it promotes the deposition of calcium in the bones.

The need for calcium is especially high for infants, children, adolescents, pregnant women, and nursing mothers. Milk is the best dietary source of calcium. Dried peas and beans are another good source. Cheese, too, is a source of calcium, although some of the calcium stays in the whey during the cheese-making process. People who eat a lot of spinach or whole cereal grains, especially wheat, do not receive the full benefit from the calcium in their diet because these foods contain oxalic and phytic acid which combine with the calcium and make it unavailable for use by the body. Diets high in purified protein alter the metabolism of calcium and reduce the amount available.

Phosphorus. Phosphorus is also a constituent of teeth and bones. It is part of an important class of molecules that transfer and store energy in many metabolic reactions. As such, phosphorus is part of the molecule that supplies energy for muscle contraction and for the conduction of impulses along nerves. Cyclic AMP, a derivative of the nucleotide phosphates, functions as a mediator in hormonal action. Most of the neuromuscular and hormonal control and integration systems in the body depend on molecules that contain phosphorus. Phosphorus levels are regulated by some of the same control systems that regulate calcium. Phosphorus is present in nearly all foods, and deficiency is rare.

Potassium, Sodium, and Chloride. Potassium and sodium both have a role in the conduction of nervous impulses. Like calcium, these minerals are essential for maintaining normal heart rhythms. The concentration of dissolved sodium in body fluids has a significant effect on the balance between the fluids in the blood plasma and the fluids that bathe the cells. People who suffer from high blood pressure are advised to limit their intake of sodium. Eliminating table salt is a first step. However, many processed foods have so much sodium that even when table salt and cooking salt are eliminated, the average salt consumption is still well above the nutritional requirement. In the United States, in fact, the average salt intake is 20 times greater than the nutritional requirement.

When we eat table salt, which is sodium chloride, we, of course, ingest another essential nutrient, the chloride ion. The chloride ion is one of the negatively charged ions. It is responsible for maintaining fluid balance, and it is a component of gastric juice.

Magnesium. Magnesium is another mineral that is dissolved in the blood and tissue fluids in ionic form. Like calcium, potassium, and sodium, magnesium must be present to assure the integrity of cell membranes. Magnesium also functions in many enzyme systems. Magnesium is plentiful in foods, and deficiency is rare in the general population. Alcoholics and people with absorption disorders, however, often suffer from magnesium deficiencies.

Iron. There is another group of minerals that must be present in the diet, but not in such great quantities as those already discussed. This group includes iron, manganese, and iodine. Iron has two principal functions: it is part of the hemoglobin molecule which binds oxygen and carries it in the red blood cell, and it is a cofactor in some enzymes. The body can store iron in the liver, spleen, and kidney. Infants are born with iron stores, which are depleted by the age of 6 months if they have an exclusively milk diet and no supplementary iron. The amount of iron that is available in diet alone is not sufficient to maintain iron stores in pregnant women, and supplements are recommended. Many menstruating women have a marginal iron deficiency due to the blood loss from menstruation. Iron in the diet is not always fully absorbed. Absorption is enhanced by vitamin C, and it is impaired by calcium and phosphate salts, tannic acid in tea, phytic acid in whole grains, especially wheat, and by antacids. Rich dietary sources of iron are eggs, meat, poultry, fish, and liver. Whole grain and enriched breads and cereals, dried peas and beans, and dark green vegetables are also good sources of iron. Low levels of iron means that the red blood cells cannot carry as much oxygen as the body needs. Overdoses of iron tablets can be toxic, and have accounted for episodes of accidental poisoning in children.

Manganese. Manganese is part of many enzyme systems involved in protein synthesis and energy metabolism. Deficiency in the diet is rare, but symptoms include impaired reproduction, retardation of growth, congenital defects in offspring, and abnormalities in bone and cartilage.

Iodine. Iodine is an essential component of the hormone thyroxin and other thyroid hormones, and sufficient iodine is necessary for normal thyroid glands. Individuals who do not get enough iodine in the diet suffer from goiter, in which the thyroid gland becomes enlarged. Goiter can be associated with puffy skin, hair loss, obesity, reduced heart rate, lethargy and dullness. The incidence of this condition has been greatly reduced in the United States by adding iodine to table salt.

Trace Minerals. Finally, there is another class of minerals, called trace minerals, which includes zinc, cobalt, copper, chromium, selenium, and fluorine. These elements must be present in the diet, but, as the name implies, only in very small amounts. Trace minerals usually function as cofactors in metabolic reactions. Their presence is required for the activity of the enzymes that catalyze these reactions. Unless the bulk of an individual's diet consisted of highly processed, refined food, the probability of trace element deficiency is low, and supplementary sources of trace elements are not necessary.

VITAMINS: "VITAL FOR HEALTH" Vitamins got their name in the following way: A substance was isolated from rice polishings and was shown to have a curative effect for the deficiency disease beriberi. Its chemical structure was believed to be that of an amine. Since the substance was vital for health its name took *vita* (from Latin 'life') as the prefix and *amine* as the root word. Later, it was discovered that this substance and some others that were essential for good health did not have the amine structure, so the "e" was dropped from amine, and they all were given the designation vitamin. A vitamin is defined as a substance that cannot be synthesized by the organism but is required in the diet for normal health. A vitamin for some animals is not necessarily a vitamin for others. Vitamin C is a vitamin by this definition for humans but not for guinea pigs, because guinea pigs can synthesize the vitamin themselves.

The Fat-Soluble Vitamins: A, D, E, and K. Vitamins are classified on the basis of whether they are water soluble or fat soluble. The fat-soluble vitamins for humans are vitamins A, D, E, and K. Vitamin A is available in the diet in the preformed state or as a precursor of protovitamin A, which can be changed by the body after having been ingested. Active vitamin A is found in organ meats, animal fats, and egg yolks. Protovitamin A is found in green and yellow vegetables and in some fruits. Most of the active vitamin is first ingested in this form and then transformed through a series of reactions in the intestine. Vitamin A can be stored in the liver and in adipose tissue. Vitamin A has a crucial role in vision, and it also functions in growth, reproduction, maintenance of cells that line the inner and outer surfaces of the body, and stability of cell membranes. Vitamin A deficiency can be caused by insufficient vitamin in the diet or by any condition that causes a malabsorption of fat, since the vitamin itself is fat soluble. Vitamin A deficiency causes night blindness, dryness and roughness of the skin, and a drying of the cornea of the eye which can lead to ulceration and blindness. Vitamin A toxicity is possible because the vitamin is stored in fatty tissues, and stays in the body for a relatively long time. It cannot be excreted in the urine because it is not water soluble. It is impossible to tell exactly what a toxic level would be for each person since there are wide individual differences in sensitivity to high doses. The nutritional status of the person taking the vitamin must be considered. If vitamin A stores in the body are already large, the threshold dose might be lower. The effects of vitamin A toxicity are acute but transient. They include a rash, dizziness, and sweating.

Vitamin D is a vitamin for humans only under certain conditions. It can be synthesized by the body from cholesterol if there is sufficient sunlight and the skin is irradiated by the sun. If there is not enough sunlight, or if the skin is always covered, then the vitamin must be supplied in the diet. A common misconception is that blacks have a high concentration of melanin in their skin to protect them from sunburn. In fact, this dark pigment prevents excessive synthesis of vitamin D from cholesterol and thus confers an evolutionary advantage to blacks living in sunny climates because it protects them against vitamin D toxicity. This protection is crucial because vitamin D is the most toxic of the vitamins in excessive amounts. Toxic levels of vitamin D in the body cause deposition of calcium in internal organs such as the kidney, in arteries, and in soft tissues. Vitamin D occurs naturally in animal foods such as eggs, liver, and butter, and in fatty fish. It is sometimes added to milk. Vitamin D is synthesized in the kidney and after synthesis leaves the kidney and goes to receptor proteins in intestinal cells. It enters the cell nucleus where it directs the synthesis of several hormones. The net effect of its action is to increase absorption of calcium and phosphorus from the gut. Vitamin D may directly influence the mineralization of bones. The effects of vitamin D deficiency are rickets in children and osteomalacia in adults.

Vitamin E is available to the human in vegetable oils, meat, egg yolks, and green vegetables. If laboratory animals deficient in vitamin E are given an antioxidant, the symptoms of vitamin E deficiency disappear. This fact has led researchers to believe that vitamin E itself functions as an antioxidant. Vitamin E is also believed to stabilize cell membranes. Deficiencies have been shown to cause infertility and reproductive failure in rats. In humans, the symptoms are increased fragility and susceptibility to bursting in red blood cells.

Vitamin K is supplied in the diet by green vegetables and can be synthesized in the intestine by bacteria. It regulates the synthesis of prothrombin and other clotting factors in the blood plasma. Infants are born with a vitamin K deficiency, and are given an injection of vitamin K at birth to protect them against the dangers of hemorrhage. If adults have a vitamin K deficiency, prothombin levels are lower, and blood clotting time is prolonged.

The Water-Soluble Vitamins: B-Complex and C. The classification of vitamins into water soluble and fat soluble has very little relationship to food sources of vitamins. It was merely a convenient terminology devised by the chemists who originally isolated and studied vitamins. However, one important consequence of water solubility is of practical concern. Water-soluble vitamins can be dissolved or leached out into the water in which foods are cooked or soaked. If this liquid is discarded, the dissolved vitamins are thrown away with it.

The B-complex vitamins are grouped together because they are functionally interrelated. All of the B-complex vitamins are coenzymes, converted through a series of reactions in the cell from the form they had when they were eaten to a biologically active coenzyme. These vitamins participate in metabolic reactions that involve energy transfer, rearrangement of molecules, and synthesis of molecules that are precursors for DNA and RNA. The B-complex vitamins have the individual names of niacin, riboflavin, pantothenic acid, pyridoxine, thiamine, biotin, folic acid, and cyanacobalamin (vitamin B_{12}).

Niacin is also known as nicotinic acid. As a coenzyme it participates in reactions in which energy is released gradually during cellular respiration. It accepts high-energy hydrogen which is passed down a series of acceptor molecules, the cytochrome system, gradually releasing energy that the cell can store or use. Humans can synthesize niacin from the amino acid tryptophan. However, because there is usually not enough tryptophan in the diet to satisfy both the vitamin and the amino acid needs, it is recommended that a source of niacin be included in the diet. Niacin is found in liver, beef, pork, whole grain cereals, fruits, and many vegetables. Niacin deficiency leads to a disease called pellagra. People who have pellagra suffer from the "three D's" of dermatitis, diarrhea, and dementia, including severe impairment of intellectual capacity and personality disintegration.

Riboflavin, or vitamin B_2, functions in chemical reactions which involve oxygen. It is present in most foods, and it is made by bacteria in the intestine (although some think that this happens after the absorption of nutrients has taken place). Especially rich sources of riboflavin are organ meats, milk products, and whole grain products. Symptoms of riboflavin deficiency are changes in the cornea, inflammation of the membranes that line the inner surface of the eyelid, disturbances of vision, and sores on the lips and tongue.

Pantothenic acid, vitamin B_3, is converted to coenzyme A. Coenzyme A plays an essential role in the oxidation of glucose and other molecules that yield energy in cellular respiration. The name for this vitamin is derived from the Greek word *pantothen* 'from everywhere.' As the name suggests it is widely available in food, especially in liver, eggs, meat, and fruits. Deficiency of this vitamin is rare, but the consequences are impairment of adrenal gland function and of antibody synthesis, and numbness and pain in toes and feet.

Thiamine, vitamin B_1, is a component of an enzyme system that releases energy when food is oxidized. Sources of thiamine are pork, the germ and outer coat of whole grains, green vegetables, fish, and milk. Thiamine deficiency leads to beriberi. This disease is prevalent among people who eat polished rice as the major constituent of their diet. It is rare in the United States except among food faddists and alcoholics. People with beriberi may suffer from muscle atrophy, paralysis, mental confusion, and congestive heart failure.

Biotin participates in reactions that add carbon dioxide to other molecules in the cell. Sources of biotin in the diet are organ meats, milk, and nutritional yeast. Deficiency in humans is rare, but the deficiency may be caused by too much raw egg white in the diet. Raw egg white contains a protein, avidin, that binds biotin and makes it unavailable. Some cases of dermatitis respond to therapy with biotin.

The coenzyme made from pyridoxine, vitamin B_6, is involved with amino acid reactions. One such reaction is deamination, which you may remember removes an amine group and prepares an amino acid to be used as an energy source in cellular respiration. Pyridoxine also takes part in reactions during which amino acids are shifted from one protein to another. Deficiency of pyridoxine is rare in the general population. Dietary sources of this vitamin are whole grains, fresh meat, organ meats, eggs, and fresh vegetables. Symptoms of pyridoxine deficiency are convulsions, depression and confusion, dermatitis, and impairment of antibody synthesis. There is clear evidence that the need for pyridoxine increases during pregnancy, and it is thought that this is due, in part, to increased protein metabolism. It is unclear at this time whether the increased need for pyridoxine can be met by the diet or whether a supplement is needed. It used to be thought that women who took oral contraceptive steroids needed a pyridoxine supplement. However, since OCS users usually do not show signs of deficiency to a greater extent than the general population, routine use of pyridoxine supplements is not currently advised. There is a high incidence of pyridoxine deficiency among alcoholics.

The coenzyme derived from folic acid has a variety of functions. Among them are energy transfer reactions, the synthesis of amino acids from precursors available in the cell, synthesis of a DNA precursor, and the formation of red blood cells. Folic acid deficiency is the most common vitamin deficiency in humans, and it is seen frequently among chronic alcoholics. During pregnancy and lactation, the need for folic acid increases and use of a supplement to the dietary sources of folic acid is recommended. Effects of folic acid deficiency are anemia, stunted growth of young infants and children, and impaired antibody synthesis. Dietary sources of folic acid are green leafy vegetables, organ meats, whole grains, and nuts.

Cobalamin has the familiar name of vitamin B_{12}. The coenzyme from this vitamin reacts with the folic acid derivatives in some of the reactions that make the precursors of DNA and RNA. The cobalamin coenzyme also helps to make a molecule that takes part in energy production via cellular respiration, and to catalyze a reaction that makes an amino acid. It plays a role in red blood cell formation, as well. The effect of vitamin B_{12} deficiency is pernicious anemia, a common affliction of older persons. Strict vegetarians also show signs of B_{12} deficiency because there are no vegetable sources of this vitamin. Good sources of vitamin B_{12} in the diet are liver, beef, pork, whole milk, eggs, and oysters.

The other water-soluble vitamin is vitamin C. Dietary sources are citrus fruits, tomatoes, potatoes, and green vegetables. Since vitamin C is heat sensitive as well as water soluble, a large amount of the potential vitamin C yield of a food can be lost by excessive cooking.

Vitamin C functions in the synthesis of collagen, a protein component of connective tissue and part of the matrix of bones. Vitamin C facilitates the absorption of iron. Vitamin C deficiency causes scurvy, a disease that used to be prevalent among sailors who had no access to fresh fruit and vegetables. Infants, food faddists, alcoholics, and the elderly sometimes have scurvy today. Infants with scurvy have defective bone formation. Other symptoms of vitamin C deficiency are swollen gums, bleeding gums, susceptibility to bruising, slow wound healing, anemia, and wasting of body tissues.

Linus Pauling, a brilliant and highly respected chemist and a double recipient of the Nobel Prize, postulated that large doses of vitamin C were helpful for preventing the common cold. Many of his colleagues believed that in proposing this Pauling had gone off the deep end. But, because of his reputation, his postulate was carefully tested in a number of clinical research trials. The usual design of the experiment was that some healthy subjects were given vitamin C and some were given a placebo, an inert substance that resembled the vitamin tablet in every way except activity. It was extremely difficult to design a convincing placebo for vitamin C because the vitamin has a characteristic taste. The frequency and severity of colds in both groups was recorded, and each group was compared at the end of the trial period. No significant differences were found between the groups. The controversy still rages. Pauling cites studies that prove his assertion. Opponents of Pauling's view claim that the results of these studies have not been reproducible, an important step in the scientific proof of an assertion. In one study the group who received the placebo (but thought they had the vitamin) had fewer colds than the group who received the vitamin. The hypothesis that vitamin C prevents colds has not been proven. During the course of the investigations of vitamin C, questions about the toxicity of the vitamin arose. Vitamin C is quite soluble in water, so it is fairly easy to excrete excess amounts in the urine. Some people believe, however, that prolonged use of vitamin C in the amount of several grams per day may irritate the gastrointestinal tract and cause diarrhea. Megadoses may cause the precipitation of uric acid and/or cystine (an amino acid) which can lead to the formation of kidney stones.

In spite of the lack of supporting evidence for vitamin C as an appropriate preventive measure for colds, many people still take the vitamin and swear by it. Why? Is it all in their heads? According to theories about the placebo effect, it may be in their heads, but the effect is the same as if the vitamin were actually acting at the cell level to prevent colds. There are many examples of subjects in double-blind clinical trials of drugs who feel better after taking a placebo. The people feel better because of a strong belief that the placebo, which they think is a drug, will make them better. In *Anatomy of an Illness,* Norman Cousins describes his experiences with megadoses of vitamin C taken during a bout with a serious collagen degenerative disease. In his case there was a reasonable basis for using vitamin C: Because the vitamin is known to have a role in collagen synthesis, it was reasonable to postulate that megadoses might reverse the course of collagen degenerative disease. Cousins postulates that his eventual recovery may have been a demonstration of the placebo effect.

Vitamin Supplements. The vitamin industry is thriving in the United States because many people practice a form of self-therapy by taking vitamin and/or mineral supplements. The basis for the feeling of well-being that many experience after taking supplements may be organic, or it may have its roots in the placebo effect. Is it necessary to take vitamins? Is it harmful? The range of foods available to the American consumer is such that anyone who eats a balanced diet and avoids highly refined food as the major source of food, probably does not need to take vitamin supplements. Vitamin deficiencies are extremely rare in normal, healthy populations. Vitamin toxicity, on the other hand, can be a problem, especially with vitamins A and D. Self-prescribed vitamins will not have a toxic effect if the daily dosage is near the Recommended Daily Allowance formulated by the Food and Nutrition Board of the National Research

Council of the National Academy of Sciences (see Table 11-2). Unfortunately, some people allow fallacious thinking to guide them when making choices about vitamins.

One fallacy is this: If some is good, then more is better. Another fallacy is assuming that if a deficiency causes one syndrome than its opposite can be cured by a vitamin. Vitamin E deficiency is known to cause sterility in male rats. Many human males take vitamin E to enhance their potency or their virility, even though potency and infertility are separate phenomena (see Chapter 4). One might do better to be guided by facts about how vitamins work when deciding how much of a vitamin to take.

You will remember that vitamins function as coenzymes in metabolic processes. As coenzymes, vitamins play a part in chemical reactions that are directed ultimately by the genes in the cell nuclei. There is a finite need for vitamins that is determined by how much of a particular enzyme the cell makes. If the genes direct a cell to synthesize 1,000 molecules of an enzyme and a vitamin dose supplies enough to synthesize 1,000,000,000 molecules of a coenzyme, the resultant holoenzyme will not be better because of the extra vitamins. The synthesis of enzymes is usually regulated by a product-inhibited negative feedback system. This means that when the con-

TABLE 11–2 The Recommended Dietary Allowances

Nutrient	*Children (1–10)*	*Males (23–50)*	*Females (23–50)*	*Pregnant*	*Nursing*
Protein (grams)	23–36	56	46	76	66
Vitamin A (IU)	2000–3000	5000	4000	5000	6000
Vitamin D (IU)	400	400	400	400	400
Vitamin E (IU)	7–10	15	12	15	15
Vitamin C (mg)	40	45	45	60	80
Folic acid (ug)	100–300	400	400	800	600
Niacin (mg)	9–16	18	13	15	17
Riboflavin (mg)	0.8–1.2	1.6	1.2	1.5	1.7
Thiamine (mg)	0.7–1.2	1.4	1.0	1.3	1.3
Vitamin B_6 (mg)	0.6–1.2	2.0	2.0	2.5	2.5
Vitamin B_{12} (ug)	1.0–2.0	3.0	3.0	4.0	4.0
Calcium (mg)	800	800	800	1200	1200
Phosphorus (mg)	800	800	800	1200	1200
Iodine (ug)	60–110	130	100	125	150
Iron (mg)	10	10	18	36	18
Magnesium (mg)	150–250	350	300	450	450
Zinc (mg)	10	15	15	20	25

IU = International Units; mg = milligrams = 1/1000 gram or .001 gram; ug = microgram = 1/100,000 gram or .000001 gram.

Source: "Food and Nutrition Board: Recommended Dietary Allowances," 8th ed. Publication Number 1694. Washington, D.C.: National Academy of Sciences-National Research Council, 1974.

centration of enzyme reaches a certain level the synthesis mechanism is turned off. The extra vitamin is then excreted if it is water soluble or it is stored in the body if it is fat soluble. If large doses of fat-soluble vitamins are taken over a period of time, a toxic concentration of the vitamin in the tissues may result.

Choosing Well

How can we be assured of a nutritionally adequate diet? Is it necessary to have a copy of the Recommended Dietary Allowances (RDA) constantly at hand during meal planning and preparation? No, it is not necessary, and it is probably undesirable. The Food and Nutrition Board, the group that prepared the RDA, gave wise counsel in a recent publication, *Toward Healthful Diets* (1980):

The Board expresses its concern over excessive hopes and fears in many current attitudes toward food and nutrition. Sound nutrition is not a panacea. Good food that provides appropriate proportions of nutrients should not be regarded as a poison, a medicine, or a talisman. It should be eaten and enjoyed.

How can we know if our diet provides appropriate proportions of nutrients? Nutritionists have classified the foods we eat and have established four basic food groups. They are the fruit and vegetable group, the bread and cereal group, the milk and cheese group, and the meat, fish, poultry, and beans group. It is recommended that we eat four servings daily from the first two groups, and all adults should have two servings daily from the last two groups. In addition, pregnant women should have an extra serving from the milk group, and nursing mothers should have two extra servings from the milk group. The United States Department of Agriculture has published an easy-to-use food guide that provides more detailed information about the basic four groups. It is Home and Garden Bulletin, number 228, and it is available from the U.S. Government Printing Office, Washington, D.C. 20402. The main principle to keep in mind when planning a diet is to choose from a variety of foods.

Another principle that many Americans would do well to keep in mind is to choose a diet that supplies only as many calories as daily energy needs require. Obesity is a serious health problem in the United States. About 30 percent of middle-aged women and 15 percent of middle-aged men in this country are at least 20 percent above their desirable body weight. Possible consequences of obesity include the increased risks of high blood pressure, diabetes, coronary heart disease, and gall bladder disease. The chances of death from these diseases are reduced if the obese patients lose weight. Women who are overweight are at greater risk for developing breast and uterine tumors. All obese people are higher surgical risks, and many who are obese have lower resistance to infections. While the cause of obesity differs from person to person, there are two general trends in the diet and exercise habits of Americans that predispose them toward obesity. One is high dietary fat consumption, and the other is sedentary life-style. Nutritionists are now recommending, especially for people who are overweight, the percentage of fat consumed should be reduced from current levels (which supply about 42 percent of the daily calories) to about 30–35 percent of the daily food intake. There is a group of foods that supply calories but very few nutrients: the so-called "empty calorie foods." These include sweets such as candy, sugar, jams, jellies, syrups, and sweet toppings; sweetened drinks such as soft drinks and juice drinks (as opposed to plain fruit juice); unenriched baked goods made from refined flour and sugar; beer, wine, and liquor; and nonessential fats and oils. These should be eliminated from the diet of people who are trying to lose weight and restricted in the diet of the general population.

FAD DIETS: LOOKING FOR THE EASY WAY OUT There are many fad diets that promise quick and easy weight loss through some aberrant pattern of food consumption. There are high-fat diets, high-protein diets, high-carbohydrate diets, and even high-alcohol diets. While it is tempting to contemplate effortless weight loss, the simple fact is that the only way to lose weight is to consume fewer calories than are burned by daily living. The recommended proportions of the various food classes in the diet are the following: carbohydrate, 50–55 percent; fat, 30–35 percent; and protein, 15–20 percent. Any diet that promises the dieter the freedom to eat all she wants and still lose weight is making false or misleading promises. Some diets say that we can eat unlimited amounts of certain foods that have secret "fat-melting" properties. The grapefruit was such a highly touted food a few seasons ago, there were even capsules of grapefruit extract with its "miracle weight loss" promise that one could buy. While grapefruit has much to offer, especially in terms of available vitamin C and bulk which might contribute to a sense of satiety, it is not a complete source of essential nutrients, nor does it have any weight loss inducing properties per se. Moreover, 2,500 calories worth of grapefruit put into the body of a woman who burns 2,000 calories in the course of a day will result in a gradual weight gain in that individual. Those of us who go on fad diets and starvation diets often feel bad because the foods included in these diets do not provide us with enough essential nutrients. Fortunately for our health, but unfortunately for our weight loss goals, we cannot stand to follow these diets for a prolonged time. We bounce back from starvation diets ravenously hungry and promptly regain the lost weight and sometimes even more.

PERMANENT WEIGHT CONTROL While they are certainly tedious to contemplate, the most successful diets are those that reduce an individual's weight slowly (probably the reverse of the process that led to obesity in the first place). Gradual weight loss is less harmful to health and gives the body a chance to stabilize. Although the prospect of spending the rest of one's life on a diet is grim, indeed, it is often the case that permanent changes in eating habits are necessary for permanent weight control. While no one agrees on the best source of inspiration, nor the best method of permanent weight control for each individual, many studies of weight loss programs have shown that it is easier to lose weight and keep it off if calorie restrictions are accompanied by an exercise program.

EXERCISE

Besides the obvious benefits of burning calories and easing the severity of restrictions of food intake, exercise offers many other benefits to the dieter and nondieter alike. Human beings were meant to be in motion. Although we have invented labor-saving machines that have permitted a more sedentary life-style, the minds that designed this technology are inside bodies that are adapted to a more primitive life-style—one that requires moving to gather food and hard work to assure shelter. Exercise gives our entire bodies the stimuli to which they were evolutionarily designed to respond.

Physiological Effects

What are the physiological effects of exercise? Perhaps it would be helpful to talk first about what happens in the isolated event of exercise and then discuss the effects of physical training, of "getting in shape." During exercise the heart beats faster, blood is diverted to the working muscles, and food that is stored as potential energy in the form of glycogen or fat is converted to actual mechanical energy. The cardiovascular system, the respiratory system, the nervous system, and the endocrine system are all stimulated and all function at a higher level of intensity than when we are at rest. Several hormones are secreted during exercise, and they have a variety of effects.

Effects on Hormonal Control Systems

Thyroid hormone secretion increases with these effects: the contractility of cardiac muscle increases, utilization of cholesterol from the liver is increased, and fat is mobilized from adipose tissue. Androgens are in greater concentration after exercise; both males and females have higher postexercise levels of testosterone. This causes greater nitrogen retention, facilitates protein synthesis, and stimulates glycogen storage in liver and muscles. The adrenal medulla secretes more epinephrine and the sympathetic nervous system secretes more norepinephrine. These hormones elicit the fight or flight response when an organism perceives a threat or danger, but they are secreted during exercise as well. The effects of higher levels of these hormones are rise in blood pressure, increased heart rate, decreased insulin secretion, increased blood glucose levels (because of conversion of glycogen to glucose and because of gluconeogenesis in the liver), increased oxygen consumption, liberation of fatty acids from fat stores, resistance to fatigue, dilation of blood vessels, and increased blood flow to the skeletal muscles and heart muscle. Emotionally the exerciser feels an increased sense of alertness, a sense of being ready to take on whatever comes up. Exercise also increases the level of growth hormone. This stimulates metabolic activity in the long bones. Growth hormone also acts to liberate fat from adipose tissue by increasing the synthesis of enzymes that break down fat. Growth hormone also has the effect of raising the levels of glucose in blood plasma and the levels of glycogen in the liver, perhaps by effecting gluconeogenesis. If exercising is brief or moderate, the levels of adrenal hormones stay the same or sometimes drop. If someone exercises to the point of exhaustion (from an objective, physiological viewpoint, not from a subjective feeling of being tired), the level of these hormones is raised, and a stress reaction ensues. This stress reaction is characterized by tissue breakdown, increased calcium loss in bone, depressed glucose uptake in muscles, impaired carbohydrate utilization, and increased breakdown of fats.

Training Effect

When someone has exercised over a period of time at a certain intensity (an actual physical intensity, not an emotionally perceived intensity), it is possible to experience what is called a training effect. The training effect results in physiological and psychological changes and improves the general quality of our health.

Physiological Aspects The physiological effects of training are felt throughout the body's systems. The heart gets stronger and can pump a greater volume of blood per beat. The number of capillaries in the muscles is increased. The fibers in the muscles get stronger. The number of mitochondria (places in the cell where energy-releasing reactions take place) increase in muscles. The body's capacity to absorb and transport oxygen, the aerobic capacity, is increased. The concentration of the enzymes that help to oxidize food is increased, leading to a more efficient utilization of fuel. The body composition of a trained person changes. The concentration of lean muscle mass increases, and the percentage of fat in the total body mass decreases. Persons who are trained for endurance sports have more efficient fat metabolism in their exercising skeletal muscles. The force which muscles are capable of exerting increases, and the body's capacity to do work increases.

Psychological Component There is a psychological component to the training effect as well. People who are physically fit are known to differ from people who are not trained physically in several personality categories. They show greater emotional stability, a greater sense of self-confidence and security, and lower levels of anxiety. There is speculation about the causes for these personality differences. Some say that it is personality that affects a person's decision to exercise or participate in sports in the first place; others say that exercise affects personality. Some researchers postulate that the basis for personality differences is molecular, that persons who are physically fit have different concentrations of the biochemical compounds that are believed to be related to emotional states. One group of such compounds is the endorphins, morphine-like molecules which are secreted by brain cells and believed to be a source of pleasurable and pain-masking sensations. Because there is, as yet, no direct way to measure the concentration of endorphins in live human subjects, the status of the assertion that endorphins are the source of good feelings among physically fit people is still highly speculative. It is known that as physical fitness increases, depression, anxiety, and self-centeredness decrease. Many mental health professionals have discovered that running or long distance walking are effective treatments for depressed patients. Sometimes, if a person *thinks* that she is physically fit there is an enhanced sense of self-esteem, even if there is no accompanying physiological evidence of increased fitness. And conversely, people who are physically fit but are not aware of being fit do not have as high a sense of self-esteem as people who think they are fit and get some social feedback about their fitness. There is a correlation between body image and sense of self.

Improved Health Status Another effect of physical training is improved health status. Gradually increasing exercise has been shown to be an important part of rehabilitation programs for people who have suffered heart attacks. Non-insulin-dependent diabetes can be helped by exercise. Researchers have discovered a marked facilitation of formerly masked insulin receptor sites when obese, non-insulin-dependent diabetic patients participate in exercise programs. People who are physically fit are more resistant to infections, and they have a longer life expectancy than people who are not fit.

Figure 11-1

Physical Fitness for All

ATTAINING FITNESS Given the benefits of physical fitness, how may they be attained? Is physical fitness possible for anyone, or are there some persons who are beyond help? In theory, anyone may become physically fit. Women and men respond to physical training in the same way with one exception: men have the potential to develop greater muscle mass than women. Although older people respond more gradually than younger people to physical training, their response to physical training is not impaired with age. Older people who have become incapacitated by muscle weakness or by stiffness have experienced significant increases in mobility after participating in exercise programs designed especially for their needs. The effects of years of sedentary living have been counteracted by regular exercise. There are some differences in the degree of an individual's response to physical conditioning, but not in the kind of response that the body can make. It is recommended that anyone over age 35 or anyone with known serious health problems consult a doctor before embarking on a conditioning regimen. A key principle is to proceed gradually. There is a threshold at which the training effect is achieved, and there is a limit beyond which increased exercise will have no more results. Up to that limit, it is possible to train the body to higher levels of fitness by gradually increasing the amount of exercise that one does. Attaining fitness is essentially a stimulus-response phenomenon. The exercise places a mild stress on the body, and the body responds to that stress adaptively. If someone proceeds too rapidly in a conditioning program, or if one goes beyond the innate limits of the body, injuries can result. An easy way to tell where the limits are is to be aware of pain. It is advised to exercise to the point of pain, but to stop at that point.

Giving up on Exercise A sad fact for those who would like to become fit and then give up exercising is that the body is governed by a "use it or lose it" principle as far as fitness is concerned. When the stimulus ceases the response ceases. If exercise is interrupted the participant regresses to previous fitness levels. A reduction in the work capacity of the muscles occurs in about two weeks after the training stops, and the muscle strength reverts to pretraining levels within ten weeks to eight months after detraining. Cardiorespiratory fitness is reduced by half in four to twelve weeks after detraining. One way to assure continuity in an exercise program is to choose an activity suited to your personality. Some studies show that people who exercise first thing in the morning are more likely to stick with an exercise program. Taking care to go slowly and avoid injuries while working up to a level of fitness is also important. Like the diets that promise easy weight loss, there are exercise programs that proclaim the possibility of total fitness in 30 minutes a week.

A Realistic Exercise Program The American College of Sports Medicine offers a more realistic guide for those who are wondering about the kind of fitness program to choose. They recommend that three to five days per week be devoted to training sessions. They advise an exercise intensity that calls upon 60–80 percent of the heart rate reserve or 50–85 percent of the maximum oxygen uptake potential. The simplest way to determine if you have approached 60–80 percent of your heart rate reserve is to take your pulse right after the exercise you have chosen. A rough approximation is that 135–150 beats per minute corresponds to the recommended heart rate reserve percentage. Fifteen minutes to one hour of continuous activity are desirable. Higher intensity exercise requires shorter duration; lower intensity exercise requires longer duration. Nonathletic adults are advised to choose lower intensity, longer duration activities. The activities that are recommended use large muscle groups, can be maintained continuously, and are rhythmic. Such activities are running, jogging, walking, hiking, swimming, skating, bicycling, rowing, cross-country skiing, and rope skipping. While there are other physical activities that are pleasurable and beneficial, the activities recommended by the American College of Sports Medicine are those which promote optimal cardiovascular fitness.

Women and Exercise: Choosing and Doing

For many years women were excluded from participation in exercise and sports, if not by explicit prohibition at least by unsupportive attitudes. There was considerable conflict between the feminine role assigned to women by society and the perceived attributes of someone who exercised: "One of the oldest and most persistent folk myths, and one of the main deterrents to female sports participation, has been the notion that vigorous physical activity tends to masculinize girls and women" (Sage and Loudermilk, 1979). Some forms of exercise for women were condoned for cosmetic reasons, but active participation in sports was discouraged.

There is no basis in fact for the myth that exercise masculinizes women. One has only to recall the discussion in Chapter 2 which explains the process of sexual differentiation to realize that an environmental force cannot change what is genetically determined. Although behavior patterns exhibited by women when they exercise may seem masculine when judged by certain social norms, the fact is that exercise does not masculinize females. Fortunately, the social barriers to women's participation in exercise and sports are disappearing and it is becoming easier for every woman to enjoy the benefits offered by exercise if she chooses it.

Women and men respond to physical conditioning in the same way, but there are two uniquely feminine experiences that are sometimes talked about with respect to exercise: menstruation and pregnancy. Some may wonder if exercise has any effects on menstruation, or if menstruation has any effects on sports performance. Data relevant to these issues have been analyzed extensively with the following findings: Menarche has been delayed in some girls who train intensively for endurance sports and, in some cases, among girls who train for ballet. The theoretical basis for this is that the training causes a reduction in fat reserves in the body. While this theory is still in dispute, it holds that fat levels below 12–14 percent of total body mass cause delayed menarche. In the same way, women who are highly trained and have reduced fat stores experience menstrual suppression or menstrual irregularities. In all cases, when body fat levels rise after training ceases, menstruation recommences.

Another myth that pervades many areas of female endeavor is that performance is impaired at or near the time of the menstrual period. An analysis of the performances of female athletes in Olympic events showed that many world records were broken by menstruating women. Self-reports about performance among female athletes show that few women believe their performances to be impaired by menstruation. On the other hand, many female runners report that running is a helpful remedy for premenstrual tension or for the pelvic congestion that some women may experience around the time of the menstrual period.

Finally, there is the question about pregnancy and exercise. Most obstetricians believe that a woman can continue any physical activity that she was accustomed to before pregnancy throughout the course of a normal pregnancy. If a woman has a history of miscarriages, however, she may be advised to limit her activity. Regular exercise may make a pregnancy more comfortable. Some people believe that women who are physically fit have an easier delivery.

Others believe that the sensory stimulation received by the fetus carried by a physically active mother is beneficial.

There is nothing in the biological or social experience of being female today that prevents a woman from experiencing the advantages of being physically fit. It's all there for the choosing ... and the doing.

REFERENCES

American College of Sports Medicine. "The Recommended Quantity and Quality of Exercise for Developing and Maintaining Fitness in Healthy Adults," *Med. Sci. Sports* 10, no. 3 (Autumn 1978): vii–x.

Atomi, Y.; K. Ito; H. Iwasaki; and M. Miyashita. "Effects of Intensity and Frequency of Training on Aerobic Work Capacity of Young Females," *Journal of Sports Medicine* 18, no. 3 (March 1978): 3–9.

Benson, Herbert. *The Mind/Body Effect.* New York: Simon and Schuster, 1979.

Bray, G. A. "The Overweight Patient," *Adv. Internal Medicine* 21 (1976): 267–308.

Cousins, Norman. *Anatomy of an Illness.* New York: W. W. Norton, 1979.

Cox, J. S., and H. W. Lenz. "Women in Sports: the Naval Academy Experience," *American Journal of Sports Medicine* 7, no. 6 (November-December 1979): 355–57.

Dahlkoetter, J.; E. J. Callahan; and J. Linton. "Obesity and the Unbalanced Energy Equation: Exercise vs. Eating Habit Change," *J. Consult. Clin. Psychol.* 47, no. 5 (October 1979): 898–905.

Dale, E.; D. H. Gerlach; and A. L. Withite. "Menstrual Dysfunction in Distance Runners," *Obstet. Gynecol.* 54, no. 1 (July 1979): 47–53.

Food and Nutrition Board, National Research Council. *Recommended Dietary Allowances,* 9th ed. Washington, D.C.: National Academy of Sciences, 1980.

______. *Toward Healthful Diets.* Washington, D.C.: National Academy of Sciences, 1980.

Frisch, R. E. "Fatness and the Onset and Maintenance of the Menstrual Cycle," *Res. Reprod.* 6, no. 1 (1977): 1.

Heaps, Roger A. "Relating Physical and Psychological Fitness: A Psychological Point of View," *Journal of Sports Medicine* 18, no. 4 (December 1978): 399–408.

Ingjer, Frank. "Maximal Aerobic Power Related to Capillary Supply of the Quadriceps Femoris Muscle in Man," *Acta Physiol. Scand.* 104, no. 2 (October 1978): 238–40.

Koivisto, V., A. Soman, P. Felig. "Effects of Acute Exercise on Insulin Binding in Obesity," *Metabolism* 29, no. 2 (February 1980): 168–72.

Lappé, Frances Moore. *Diet for a Small Planet.* New York: Ballantine, 1976.

Milvy, Paul, ed. "The Marathon: Physiological, Medical, Epidemiological, and Psychological Studies," *Annals of N.Y. Academy of Sciences,* Vol. 301 (1977).

Pauling, Linus. *Vitamin C, the Common Cold, and the Flu.* San Francisco: W. H. Freeman, 1976.

Rombauer, Irma S., and Marion Rombauer Becker. *The Joy of Cooking,* Indianapolis: Bobbs-Merrill, 1964.

Rosenberg, Magda. *Sixty-Plus and Fit Again.* Boston: G. K. Hall & Co., 1978.

U.S. Department of Agriculture, Science and Education Administration. *Food,* Home and Garden Bulletin No. 228. Washington: USDA, 1979.

Index